SpringerBriefs in Geography

T0184914

For further volumes:
http://www.springer.com/series/10050

Alan Wilson

The Science of Cities and Regions

Lectures on Mathematical Model Design

 Springer

Alan Wilson
Centre for Advanced Spatial Analysis
University College London
Gower Street
London
WC1E 6BT
UK
e-mail: a.g.wilson@ucl.ac.uk

ISSN 2211-4165   e-ISSN 2211-4173
ISBN 978-94-007-2265-1   e-ISBN 978-94-007-2266-8
DOI 10.1007/978-94-007-2266-8
Springer Dordrecht Heidelberg London New York

Library of Congress Control Number: 2011938799

Printed on acid-free paper

Springer is part of Springer Science+Business Media (www.springer.com)

Preface

The study of cities and regions—understanding their workings and evolution—is one of the grand challenges of twenty-first century science. As with much of science, mathematical modelling provides a significant contribution to this understanding. There have been major developments in this field over a fifty year period and indeed historical precedents back into the nineteenth century and indeed earlier. The provenance of this book is a series of lectures given in University College London in 2010–2011. Their objective is twofold: first, to provide an elementary introduction to the field, drawing attention to significant developments in the history of urban and regional modelling that are of contemporary relevance; and secondly, to draw the audience into the contemporary research agenda. To meet these objectives, the book has a relatively unusual structure. In the early chapters some of the iconic models of earlier eras are presented both to illustrate and fix ideas and because their underlying ideas have an ongoing role in current developments. For example, Lowry's 'Model of metropolis' is presented in Chap. 1, the retail model as a key demonstrator in Chap. 2 and variants of the Lotka-Volterra prey-predator model, embracing a spectrum from ecology to political science, in Chap. 3.

Models are representations of theories and the interdisciplinary nature of these underpinnings are outlined in Chap. 4. The following three Chaps. 5, 6 and 7— begin to add more depth so that the reader is equipped to approach the research challenges that are articulated in Chap. 9. Chapter 8 is an interlude that shows how urban and regional modelling is a part of the broader and now fashionable field of complexity science.

Because this material arises from a course of lectures, the book is inevitably short. I have sought to compensate for this by offering further reading and an extensive bibliography which connects both to the history and to contemporary research.

London, June 2012   Alan Wilson

Acknowledgments

I am grateful to a number of people for help with, indeed supply of, various figures: Martin Clarke, University of Leeds, for Fig. 1.2; Alex Singleton, University of Liverpool and Oliver O'Brien, University College London, for the data for Fig. 1.3 and Joel Dearden, University College London for drawing it; Joel Dearden for Figs. 2.2, 2.3, 2.4, 2.5, 7.1, 7.2, 7.3, 7.4, 7.5, 7.6, 7.7, 7.8 and 9.1, and Francesca Pagliara for Fig. 9.2. I would like to thank a number of publishers for giving permission to reproduce a number of these figures that have been previously published: Springer for Figs. 1.2b, 2.3, 2.4 and 2.5; Prentice Hall for Figs. 4.2, 4.3 and 4.4; Edward Arnold for Fig. 5.1; and Histoire et Mésure for Fig. 5.3. I am happy to acknowledge the US National Oceanic and Atmospheric Administration for Fig. 7.3.

Contents

Chapter 1
Models and Systems: The Lowry Model as an Example

1.1 Models and Their Uses

A 'science of cities and regions' seeks an in-depth understanding of their workings and how they evolve over time. This is not only interesting as science, but is also potentially useful in a variety of planning contexts. In common with many other sciences, mathematical modelling provides a valuable approach. There is a history of 50 years or more of serious development and therefore a substantial body of literature and ideas. It is the purpose of these lectures not to review this vast body of work systematically, but rather to elucidate some of the basic principles of model design and to indicate how they can be applied. This can be thought of as the beginnings of assembling a tool kit for model building.

Models can be thought of as representations of *theories* of *systems of interest*. They offer insight and understanding and potentially provide 'What if?' forecasting capabilities for planning and for public policy development. They can be the 'flight simulators' for urban and regional planners whether in the public or commercial sectors.

The argument can be usefully structured in terms of:

*s*ystems of interest, and representations of systems;
*t*heory; and
*m*ethods for *m*odel building.

Hence, the idea of the 'STM approach': define a system of interest; decide what theory contributes to the understanding of the system; dip into the tool kit to find appropriate methods for model building. This structure will show that there are many possible ways of approaching model building—hence the idea of a tool kit.

We follow the STM sequence, more or less, to build the argument that follows. We focus first on systems but we illustrate the argument by introducing examples of models at an early stage. These illustrative models also provide an historical context. In the rest of this introductory chapter, we outline ways of representing urban and regional systems in mathematical terms and to fix ideas, we present one

A. Wilson, *The Science of Cities and Regions*, SpringerBriefs in Geography, DOI: 10.1007/978-94-007-2266-8_1, © The Author(s) 2012

of the iconic models—that of I. S. Lowry from his famous paper of 1964. In Chap. 2, we outline the retail model to show how another iconic model of the 1960s is still highly relevant today. In Chap. 3, we turn, again as examples, to models that have their roots in ecology and political science—which are interesting in themselves— but also contribute to the development of dynamic urban and regional models. This completes the introduction to systems, illustrated through some of the core models.

We then review the concepts that underpin modelling in Chap. 4—the 'theory' component of STM—and in Chap. 5, we review the methods that are customarily used to deploy these concepts in models. In Chaps. 6 and 7, we then discuss in turn spatial interaction models and dynamic models of system evolution in more depth. Urban and regional models can now be seen as a substantial component of what has become know as complexity science and these connections are outlined in Chap. 8. In Chap. 9, we jump to the research frontier and outline a number of challenges for the future. Our ambition, therefore, is to progress from an intro-duction of the main ideas and iconic modelling to a charting of the ongoing research agenda.

1.2 Systems of Interest and Their Representation

1.2.1 System Representation

The main elements of a city or region are the population, a set of organisations, and the (infra)structures—the buildings and, for example, the transport, commu-nications and utilities networks. We have to locate each of these in space and time. There is always an immediate issue, perhaps often a difficulty, in deciding the appropriate levels of aggregation for system description. In an obvious sense, the finer the level of detail, potentially the better; but this has to be set against fea-sibility and, often, data availability. We have to decide whether characteristics— say the income distribution of a population, should be represented in discrete groups or should be continuous. Similarly, should space and time be treated as discrete or continuous? We will see later that this kind of preliminary design decision has significant implications for the available mathematical tools.

1.2.2 Urban and Regional Systems

It is useful to keep in mind the diagram shown in Fig. 1.1 as we seek to develop system descriptions. This shows the elements in broad terms—the population and the economy as a framework with the addition of activities, interactions and infrastructure—all by location.

People reside, work, shop and use services at a variety of locations. Indeed, representing spatial interaction is a key underpinning for many urban and regional models. These interactions are carried on networks which can be thought of in

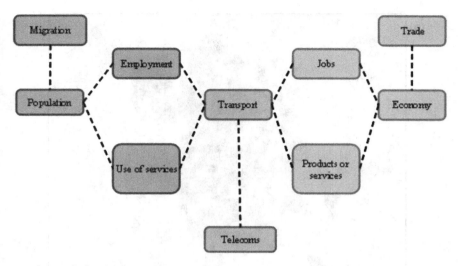

Fig. 1.1 The structure of an urban and regional system

terms of the real—transport or communications—or the symbolic–the topological. It will be necessary to identify routes through networks and to devise algorithms for the assignment of flows to network links.

In general, it will be assumed here that space is represented by a discrete zone system. Such a system may consist of official units such as wards or census districts, or of a grid. Our system elements and their activities can then be located on the zone system and interactions as flows between zones. In the latter case, it is useful to think of some zones as 'origin zones', or sources; and some as 'destinations zones', or sinks (Fig. 1.2).

Once we have all the definitions in place, we can build a 'picture' of any particular system of interest at a point in time; and then its development and evolution through a sequence of points in time. With effective search, it becomes possible to build a model based e-atlas—a topic to which we will return later.

The potential of an e-atlas can be seen through another example of a zoning system, this time showing university students whose residential base is London. In this case, geodemographic analysis has been used to characterise students as well as place of residence. In Fig. 1.3, we show the university destinations of students from well-off households and then those from 'blue-collar' households in inner London. The differences are striking! (See Singleton et al. 2011 for more detail).

1.2.3 The Lowry Model

To fix ideas, it is appropriate to introduce a model as an example at this early stage and Lowry's 'Model of metropolis' fits the bill very well. I. S. 'Jack' Lowry was working for the RAND Corporation in Santa Monica when he was assigned the

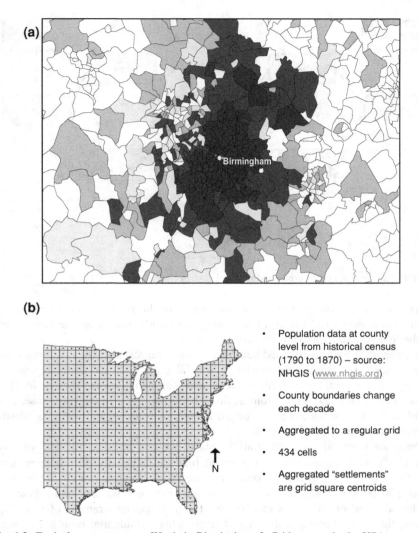

Fig. 1.2 Typical zone systems. **a** Wards in Birmingham. **b** Grid squares in the USA

task of modelling the Pittsburgh region. By the mid 1960s, there were highly developed models of urban transport systems—of which more in Chap. 4 below. It was realised at an early stage that transport investment affected land use and vice versa, so what was needed was not just a transport model but a comprehensive urban model that incorporated land use-transport interaction. There were early attempts to build such a model in Chicago and Philadelphia but it was Lowry's model of Pittsburgh that pointed the way for much future development. His model is simple—deceptively—so elegant and truly comprehensive. For our present

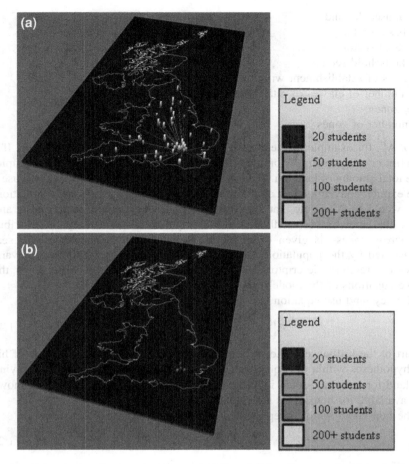

Fig. 1.3 University destinations for London-based students. **a** From 'well-off' backgrounds, Greater London. **b** From 'blue-collar' backgrounds, inner London

purposes, it shows how to define the variables that describe a city and how to assemble the elements of the model.

Lowry's main variables are

A = area of land
E = employment
P = population
c = trip cost
Z = constraints

to which should be added the following to be used as subscripts or superscripts:

U = unuseable land
B = basic sector
R = retail sector
H = household sector
k = class of establishment within a sector
m = number of classes of retail establishment
i, j = zones
n = number of zones

So, A_i^H, for example, is the area of land in zone i that is used for housing. If a subscript or a superscript is omitted, this implies summation. So A_i, for example, is the total amount of land in i. (This is very economical, though later we will use a more explicit convention that an asterisk replacing an index represents summation, so A_i would be A_i^*.) Note that there are two kinds of economic sector: basic and retail—the latter further subdivided. Basic employment—and its spatial distribution across zones—is given exogenously. Retail employment, as we will see, is generated by the population. Once this simple principle of building the variables—the region's descriptors—is understood, we can easily write down the twelve equations of the model.

The key land use equation is

$$A_j^H = A_j - A_j^U - A_j^B - A_j^R \qquad (1.1)$$

Part of the brilliance of Lowry's model was the way he captured some of his key hypotheses within the equations. In this first equation, he is in effect saying that land for basic and retail industries can always outbid housing, so this shows land available for housing as a residual.

The household sector is represented by the following set of equations:

$$P = f\Sigma_j E_j \qquad (1.2)$$

$$P_j = g\Sigma_i E_i f_{res}(c_{ij}) \qquad (1.3)$$

$$\Sigma_j P_j = P \qquad (1.4)$$

$$P_j \leq z^H A_j^H \qquad (1.5)$$

This sequence generates the population from employment and begins the process of housing them. The first (1.2) calculates total population as proportional to total employment. The second (1.3) allocates this population to zones, i. $f_{res}(c_{ij})$ is a declining function of travel cost from i to j, thus building in the likelihood that workers live nearer to their workplace. The third Eq. 1.4 enables g in (1.3) to be calculated as a normalising factor. The fourth equation is particularly interesting and also shows how the model is more complicated than appears at first sight. z^H is the unit amount of land used for residences and so this equation is constraining the numbers assigned to zone i in relation to land availability. This is one of the subtleties—and part of the trickiness—of the model: the equations have to be solved iteratively to ensure that this constraint is satisfied.

The retail sector is represented by

$$E^{Rk} = a^k P \tag{1.6}$$

$$E_j^{Rk} = b^k [c^k \Sigma_i P_i f^k (c_{ij}) + d^k E_j] \tag{1.7}$$

$$\Sigma_j E_j^{Rk} = E^{Rk} \tag{1.8}$$

$$E_j^{Rk} > z^{Rk} \tag{1.9}$$

$$A_j^R = \Sigma_k e^k E_j^{Rk} \tag{1.10}$$

$$A_j^R \le A_j - A_j^U - A_j^B \tag{1.11}$$

These six equations determine the amount of employment generated in the retail sector. The total in sector k within retail is given by the first Eq. 1.6, and this is spatially distributed through the second, (1.7). As with the residential location equation, the function $f^k(c_{ij})$ is a decreasing function of travel cost, indicating that retail facilities will be demanded relatively nearer to residences. c^k converts these units into employment. The term $d^k E_j$ represents use of retail facilities from the workplace. b^k is a normalising factor which can be determined from Eq. 1.8. Equation 1.9 imposes a minimum size for retail sector k at a location (No school for half a dozen pupils for example!). Equations 1.10 and 1.11 sort out retail land use, the first calculating a total from a sum of k-sector uses—e^k converting employment into land—and the second specifying the maximum amount of retail land—in effect giving 'basic' (which has been given exogenously) priority over retail. In this case, unlike the residential case where P_i was constrained by land availability, retail employment is not so constrained. Lowry argued that, if necessary, retail could 'build upwards'. If A_j^R from (1.10) exceeds $A_j - A_j^U - A_j^B$, it is reset to this maximum, but employment does not change.

Total employment is then given by

$$E_j = E_j^B + \Sigma_k E_j^{Rk} \tag{1.12}$$

This final equation simply adds up the total employment in each zone. The equations are solved iteratively, starting with $E_j^{RK} = 0$. As we have noted, the model is quite sophisticated in the way it uses constraints to handle land use and it is also a useful illustration of something we need always to bear in mind in model design—the distinction between exogenous and endogenous variables. In this case, the given location of basic employment is the exogenous driver but the $\{c_{ij}\}$ array can also be seen as reflecting the (exogenously specified) investment in transport. As we will see later, progress in modelling involves making more of the exogenous variables endogenous.

The Lowry model became iconic because it represented the main ideas that would underpin any comprehensive urban model in the simplest possible way. It could then be progressively refined and this is what has happened over subsequent decades.

1.2.4 Formal System Description and Accounts

Even at this early stage, it is worth showing how we can be more formal (and abstract) in describing a system of interest. We can begin with *activity* variables, which we can label, generically, as the XYZ-variables—X being some kind of origin (or production total) and Z being a destination (or consumption) total. Y represents the interaction, the flow of 'products' from one zone to another (The 'products' can include people as labour, for example, so this is quite a general idea.). More explicitly, these can be written as production (origin totals), interaction and consumption (destination totals) arrays: $\{X^{mg}_{i(m)}\}, \{Y^{mng}_{i(m)j(n)}\}$ and $\{Z^{ng}_{j(n)}\}$ where m and n are sectors and g is a good or service produced in sector m and used in sector n. (Spatial) prices can be associated with each g, such as p^g_i. The zonal subscripts have '(m)' and '(n)' added to indicate that they may be different systems—though usually that understanding would be implicit.

To fix ideas, consider a simplified version of this structure in which we drop the 'type of good' (g) so that X^m_i is total production by sector m in i and Z^n_j is the consumption of this in sector n in j. Y^{mn}_{ij} is the flow. Since the X-elements are total products, this array is sometimes known, for obvious reasons, as the make matrix; the Z-elements, as totals used, then form the absorption matrix.

We can show how to present these arrays in different kinds of table and this begins to offer insights on how the notation can be used to reveal the workings and structure of our system of interest. We begin with Fig. 1.4, the simplest plot, with sector activities against zones.

Again to fix ideas, we have defined eight sectors—population, agriculture, resource, manufacturing, retail, public service, residential and land. (It is easy to see how they would map onto Lowry's categorisation: the first and seventh as 'residential', the second, third and fourth as 'basic' and the fifth and sixth as 'retail'.) It was Lowry who used this table to categorise types of model and we gain some insight into the nature of the model building task from this perspective. The rows represent the spatial distribution of a sector; the columns, land use in a zone. Some models focus on one or other of these perspectives. A challenge is to retain both.

Figure 1.5 shows inter-sectoral flows.

As we will see in Chap. 4, this will be the basis of an input–output model.

Finally, we can combine these two tables and represent Inter-sectoral interzonal flows as in Fig. 1.6.

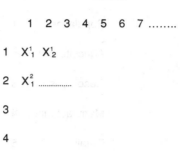

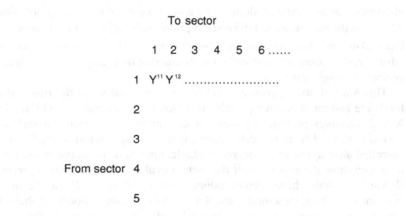

Fig. 1.4 Activities versus zones

Fig. 1.5 Inter-sectoral flows

We will use this later as the basis of a multi-regional input–output model. Each cell in this full representation represents a flow from a sector in one location ('make') to a sector in another ('use')—labelled Y_{ij}^{mn}. If we can find appropriate units—people (jobs), goods of a certain type, money (via prices)—these flow

	Sector	Structure	Flow	Product
Fig. 1.6 Inter-sectoral inter-zonal flows				
Population	1	W^1_i	$\{Y^{1n}_{ij}\}$	$\{X^1_i\}$
Agriculture	2	W^2_i	$\{Y^{2n}_{ij}\}$	$\{X^2_i\}$
Resources	3			
Manufacturing	4			
Retail	5			
Public services	6			
Residential	7			
Land	8			
Use			$\{Z^n_j\}$	

elements can be summed along rows and down columns. The production totals $\{X^m_i\}$ are the row sums and the consumption totals $\{Z^n_j\}$ are the column sums. We have also added a column of W-variables which are intended to represent 'structures'—essentially the system infrastructure that 'carry' the activities—retail centres, housing and so on.

This kind of array becomes a set of accounts; and when the row and column totals are known, it is often possible to estimate the interactions with models. The X-Y-Z variables provide a picture of the functioning system through a (short) period in time; if there is a disturbance, there is a rapid return to equilibrium—the so-called *fast dynamics*. The more challenging task is to understand the *slow dynamics*–how the structures of the system evolve—and these are represented the W-variables which have been introduced into the table. The X and Z arrays can be functions of these structural variables (and/or Y-sums). Some of these will be specified exogenously, some can be predicted and some planned. The assignment of structural variables to these categories determines the type of model and this will be a major design decision. We can think of this table as a cross section in time—which can then be thought of as one of a series of such cross sections. The potential structural 'slow dynamics' include changes in demographics, infrastructure and technologies and new transport systems.

We end by noting two further possible refinements that can make the system representation more realistic. First, as already noted, different sectors could have different zone systems, so we might say that i(m) and j(n) are the typical zones for sectors m and n. In practice, however, even when this is the case—as for example for residential areas and retail centres as in Fig. 1.2a or in Fig. 1.3 for university students—the extra m and n labels are usually dropped and 'understood'. Secondly, we should note that the m, n and g superscripts can themselves all be lists if more detail is required. If, for example, m represented the population sector, then we might want to further subdivide it by categories such as age, gender, income and so on. This makes this notation easily extendable.

It can be shown that any of the usual models can be represented in this framework—spatial demographic and economic models, spatial interaction models and so on—by appropriate definitions of 'sectors'. If this system can be specified, and appropriate data associated with it, then this is a basic systems description. It also becomes a way of organising a GIS—a subject to which we will return later.

We will now proceed to further examples of models, in each case, at this stage reverting to more specific notations. However, understanding the variety of possible descriptions through algebraic arrays is a very important part of model design.

Reference

Singleton AD, Wilson AG, O'Brien O (2011) Geodemographics and spatial interaction: an integrated model for higher education. J Geog Syst. DOI:10.1007/s10109.010-0141-5

Further Reading

Goldner W (1971) The Lowry model heritage. J Am Inst of Planners 37:100–110
Lowry IS (1964) A model of metropolis, RM-4035-RC. The Rand Corporation, Santa Monica
Lowry IS (1967) Seven models of urban development: a structural comparison. The Rand Corporation, Santa Monica
Wilson AG (1971) Generalising the Lowry model. London Pap Regional Sci 2:121–134

Chapter 2
The Retail Model and Its Applications

2.1 The Retail Model

The next step in the argument is through another example: a model of a retail system. This serves as a valuable demonstrator of a number of model design principles. It is both simple and easy to understand but can also be developed in a way that is rich and realistic. As usual, we have a discrete spatial system: zones for residential areas and what are taken as points for retail centres—so in this case origins, i, and destinations, j, represent different spatial systems. London is shown as an example in Fig. 2.1: there are 623 wards with centroids shown as dots, and 220 retail centres shown in blocks.

Thus if S_{ij} is the flow of money (say) spent on retail goods and services by residents of zone i in retail centre j and we denote the whole array by $\{S_{ij}\}$, in the London case, we have a 623×220 matrix. The power of modelling is demonstrated by the fact that we can write down one equation to represent the flow from i to j (S_{ij}) and the computer can simply repeat the calculation 623×220 times— which is 137,060 possible flows. We define e_i as the per capita expenditure by each of the P_i residents of zone i and W_j as a as the size of a centre, and by raising it to a power, α, it can be taken as a measure of the attractiveness of retail centre j. If $\alpha > 1$, this will represent positive returns to scale for retail centres. c_{ij} is a measure of travel cost as usual. The main variables are shown diagrammatically in Fig. 2.2.

The core spatial interaction model is then

$$S_{ij} = A_i e_i P_i W_j^{\alpha} \exp(-\beta c_{ij}) \tag{2.1}$$

where

$$A_i = \sum_k W_k^{\alpha} \exp(-\beta c_{ik}) \tag{2.2}$$

to ensure that

A. Wilson, *The Science of Cities and Regions*, SpringerBriefs in Geography, DOI: 10.1007/978-94-007-2266-8_2, © The Author(s) 2012

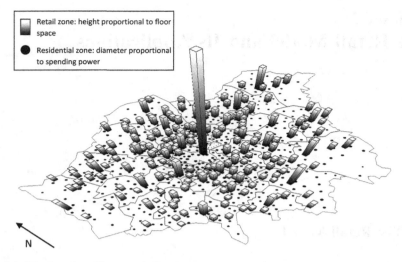

Fig. 2.1 Wards and retail centres in London

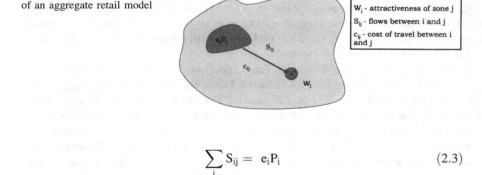

Fig. 2.2 The main variables
of an aggregate retail model

e_iP_i - demand in zone i
W_j - attractiveness of zone j
S_{ij} - flows between i and j
c_{ij} - cost of travel between i and j

$$\sum_j S_{ij} = e_i P_i \qquad (2.3)$$

That is, we build in the constraint that all the money available in i is spent somewhere. Note that in this case, we have chosen a particular declining function of c_{ij}—the negative exponential function. The reason for this, and a derivation, will be given in Chap. 6. Meanwhile, it can be taken comfortably on trust. (This is not a restrictive assumption: the exponential can be easily replaced if appropriate as we will see later.)

The constraint is on all the flows leaving a residential zone. There is no such constraint on flows entering a retail zone, and so we can use the model to calculate the total revenue attracted to a particular j. If we call this D_j, then

$$D_j = \sum_k S_{ij} \qquad (2.4)$$

which is, substituting from (2.1) and (2.2)

$$D_j = \sum_i \left[e_i P_i (W_j)^\alpha \exp(-\beta c_{ij}) / \sum_k (W_k)^\alpha \exp(-\beta c_{ik}) \right] \qquad (2.5)$$

This is a very important example because it shows how, in appropriate cases, the spatial interaction model also functions as a location model. We have already seen examples of this, of course, in the Lowry model though with more primitive interaction models.

In this presentation, we have used the customary definitions of variables, but it is a special case of the X-Y-Z-W notation of the previous chapter. $e_i P_i$ is an X-variable, S_{ij} is a Y-flow, D_j, a Z-variable and W_j is a structural variable. Indeed, the S_{ij} variables can be seen as accounts which mirror the inter-sectoral-inter-zonal table of the previous chapter—though in this simple demonstrator, there is only one sector—with $e_i P_i$ as the row sums and D_j as the column sums.

This example of a model has the advantage that it has a straightforward intuitive interpretation. If we substitute for A_i in (2.1) from (2.2), the flow model can be written in the form

$$S_{ij} = e_i P_i W_j^\alpha \exp(-\beta c_{ij}) / \sum_k W_k^\alpha \exp(-\beta c_{ik}) \qquad (2.6)$$

This shows that $W_j^\alpha \exp(-\beta c_{ij}) / \sum_k W_k^\alpha \exp(-\beta c_{ik})$ is the share of $e_i P_i$ that goes to j. This will be large if $Wj^\alpha \exp(-\beta c_{ij})$ is large compared to $\sum_k W_k^\alpha \exp(-\beta c_{ik})$ and the make up of the terms in the sum represent the *competition* of all other centres. $W_j^\alpha \exp(-\beta c_{ij})$ is a combination of pulling power (W_j^α) and the opposite effect of greater distance [$\exp(-\beta c_{ij})$].

2.2 Disaggregation

It is possible to build these models for very fine levels of detail and this is necessary to make them realistic. This has been done by consumer type, by store/retail centre type and for types of goods, so the science is well known and extensively tested. It is systematically employed—at both centre and store level—by major retailers and its value in this context is proven. Some of this experience will be described in Chap. 6. There are more challenging questions, which we will address in Chap. 9 about whether it could be applied in public sector areas.

The aggregate variables used to introduce the model could be disaggregated as follows: population by type (m)—P_i^m; type of good (g); expenditure by person type and type of good—e_i^{mg}; shopping centres by type (n). There could be different elements of 'attractiveness' by person type and type of good and so W_j could be disaggregated to become:

$$W_j^{mng} = W_j^{(1)mng} W_j^{(2)mng} \dots \dots \qquad (2.7)$$

and the cost of travel could be broken down into different elements—different kinds of time, money cost and so on to become a generalised cost:

$$c_{ij}^m = t_{ij}^m + m_{ij}^m + \ldots\ldots \tag{2.8}$$

The parameters such as α and β would also be disaggregated. For example, β^{mg} would be lower for higher value goods (g)—i.,e. generating longer trips—than for those of lower value.

2.3 Structural Dynamics

In the two examples presented so far—the Lowry model and the retail model—the model cores have been concerned with spatial interaction and the structural variables have been specified exogenously. A new challenge is to model the evolution of these structural variables. We can illustrate this with the retail example. In this case, a hypothesis for the structural dynamics can be presented as

$$\Delta W_j(t, t+1) = \varepsilon[D_j(t) - C_j(t)]W_j(t) \tag{2.9}$$

We can replace $C_j(t)$ with an assumption that costs are proportional to size—say KW_j—and then we will see in the next chapter that this is a form of Lotka-Volterra equation. The expression $[D_j(t) - C_j(t)]$, or $[D_j(t) - KW_j(t)]$ using the linearity assumption for centre costs, can be seen as a measure of 'profit' (or 'loss'). So Eq. 2.9 is representing a hypothesis that if a centre is profitable, it will grow; otherwise, it will decline. The parameter ε measures the strength of response to this signal.

At equilibrium, $\Delta W_j(t, t+1)$ is zero, so

$$D_j = C_j = KW_j \tag{2.10}$$

That is

$$\sum_i \{e_i P_i W_j^\alpha \exp(-\beta c_{ij}) / \sum_k W_k^\alpha \exp(-\beta c_{ik})\} = k_j W_j \tag{2.11}$$

These are rather fierce nonlinear equations in $\{W_j\}$ and we will explore them further in Chap. 7. Considerable progress can be made.

2.4 An Urban Systems Example

It is interesting at this stage to present a third example of a model at a different scale but which can use modified versions of the retail model equations of the previous section. We now interpret 'retail centres' as towns or cities and the flows as some composite measure of trade and migration. The variables now become:

P_i = the population of the ith city

e_i = the level of economic activity per capita—so we can distinguish in principle between 'poor' and 'rich' cities

W_i = a measure of the level of economic activity

S_{ij} = trade/migration flows

K_i = cost of maintaining the level of economic activity per unit—relates to 'rent'

c_{ij} = cost of interaction

The model to be developed—for a full description, see Wilson and Dearden (2011)—will be applied to the evolution of the North American urban system from 1790 to 1870—the period chosen because of the availability of good Census data on populations and an excellent account of the development of the railway system—in Cronon (1992). In this case, therefore, we need to add population dynamics to the system. A suitable equation, assuming that it is driven (via migration) by the economic activity—W_j-dynamics, is

$$P_i(t+1) = \mu(t)\{P_i(t)[1 + \phi_{1i}] + \phi_2 \Delta W_j(t, t+1)\} \qquad (2.12)$$

where ϕ_{1i} represents 'noise' and is a random variable less than 1; ϕ_2 measures the response of population to changes in economic activity and $\mu(t)$ is a normalising factor so that the total matches that in the available Census data.

The model is run through successive time periods—annually from 1790 to 1870. The aggregate population is increased in proportion to the known Census total and the W_js are constrained to this. We make a number of assumptions about the way the exogenous variables, represented in the vector $[\{e_i\}, \{P_i\}, \{c_{ij}\}, \{K_i\},$ $\alpha, \beta, \lambda_t]$, change over time and focus on introducing the railways exogenously to explore that effect on the urban system of cities

The system of interest is the area covering the East coast to the Midwest of the United States and we focus on the development of Chicago as the major city in the Midwest. The zone system is shown in Fig. 2.3. (This was originally seen as Fig. 1.2b in Chap. 1 but is repeated here for convenience.)

We had available population data at county level from the historical census (1790–1870)—source: NHGIS (www.nhgis.org). County boundaries change each decade and so they were aggregated to a regular grid of 434 cells. The "settlements" are then located at the grid square centroids. The transport system is represented by a spider network. This is a reasonably good approximation to a real network and is constructed by linking nearby zone centroids. We have separate links for land, water and rail. The travel cost from settlement i to settlement j is then the cost of shortest path through the spider network. When railway construction occurs the link costs change and the shortest paths are recalculated. The detail of the spider network for 1870 with land, water and railroads, is shown in Fig. 2.4.

Some results for a sample of years are shown in Fig. 2.5. The impact of the railways on the development of the Midwest is obviously of major significance.

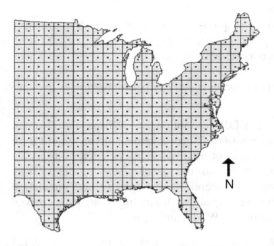

- Population data at county level from historical census (1790 to 1870) – source: NHGIS (www.nhgis.org)

- County boundaries change each decade

- Aggregated to a regular grid

- 434 cells

- Aggregated "settlements" are grid square centroids

Fig. 2.3 A grid zoning system for the North East and mid-West USA

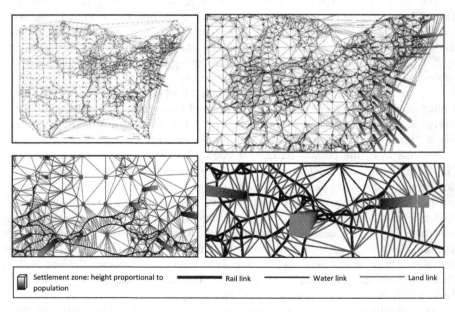

Settlement zone: height proportional to population   Rail link   Water link   Land link

Fig. 2.4 A spider network representation of the North American transport system in 1870

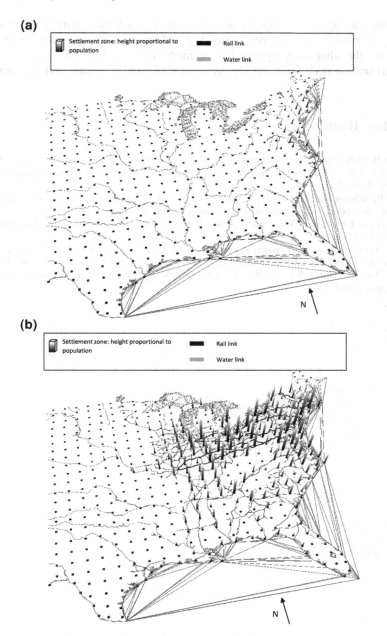

Fig. 2.5 Model-predicted growth of the North American urban system, 1790–1870

There are ongoing challenges of course. We can ask the question: what is the variety of models that can be constructed within this particular paradigm? And what are the alternative approaches to modelling this kind of system? Later, we will seek to review alternatives systematically and 'compare and contrast'.

Further Reading

Cronon W (1992) Nature's metropolis: Chicago and the great west. W. W. Norton, New York
Dearden J, Wilson AG (forthcoming-A) A framework for exploring urban retail discontinuities, Geog Anal 43:172–187
Harris B, Wilson AG (1978) Equilibrium values and dynamics of attractiveness terms in production-constrained spatial-interaction models. Env Planning A 10:371–388
Lakshmanan TR, Hansen WG (1965) A retail market potential model. J Amer Inst Planners 31:134–143
Wilson AG (2000) Complex spatial systems. Prentice Hall, Englewood Cliffs, Chapter 6
Wilson AG, Dearden J (2011) Tracking the evolution of regional DNA: the case of Chicago. In: Clarke M, Stillwell JCH (eds) Understanding population trends and processes. Springer, Berlin, pp 209–222

Chapter 3
Wars, Epidemics, Ecology and the Evolution of Spatial Structure: Connecting Models Through Generalisation

3.1 Introduction

In this chapter, we seek to develop a different kind of perspective on model design. We hinted in the previous chapter that the model of the evolution of urban structure, presented in the retail context, was similar to the Lotka-Volterra model. It will be a helpful part of our tool kit, therefore, to explore what is known about such models and how they provide valuable additions. We will also find, by exploring other fields, that from our own urban and regional perspective, we can add something to these other models. Ecologists and political scientists, for example, on the whole do not handle space well and we can facilitate progress here. It is also interesting that these kinds of models have a long history. We show how to develop the models of Richardson (1939, 1960) on arms' races and Lotka (1924) and Volterra (1938) in ecology by adding space; and we relate these to the Harris and Wilson (1978) evolutionary model of the previous chapter. A new dimension is offered by Turing's (1952) 'structure from diffusion' model as developed by Medda et al. (2009). *En route*, we collect models by Lanchester (1916), Kermack and McKendrick (1927) and Gause (1934), which both extend Richardson's warfare models and link to epidemiology.

What we see is that there have been many developments of particular versions of what is essentially the same general model over a long period with very wide applicability, but with relatively little connection between these different applications. One author who has brought all these together is Epstein through his 1997 book *Nonlinear dynamics, mathematical biology and social science* and we rely heavily on his exposition in the following presentation. We seek to build on these foundations through the addition of 'space' in a different way to that used in one instance by Epstein.

A. Wilson, *The Science of Cities and Regions*, SpringerBriefs in Geography,
DOI: 10.1007/978-94-007-2266-8_3, © The Author(s) 2012

3.2 The Richardson 'Arms Race' Model

Suppose R is the size of the red army and B of the blue one. Then consider the equations:

$$dR/dt = r - aR + cB \qquad (3.1)$$

and

$$dB/dt = b + dR - eB \qquad (3.2)$$

These hypothesise that there is a fixed rate of increase (r or b), there is attrition ($-aR$ or $-eB$) but there is an arms race: R grows as B grows and vice versa.

We have already seen that a powerful notation is valuable and so we now change notation to something that will later generalise more easily:

$$dQ_1/dt = r_1 - a_{11}Q_1 + a_{12}Q_2 \qquad (3.3)$$

$$dQ_2/dt = r_2 + a_{21}Q_1 - a_{22}Q_2 \qquad (3.4)$$

for armies of size Q_1 and Q_2 and with obvious definitions for the coefficients.

3.3 The Lotka-Volterra Prey-Predator Model

We now proceed with the general notation starting with two species. For the Lotka-Volterra prey-predator model, the equation for the prey population is

$$dQ_1/dt = (b - a_{11}Q_1 - a_{12}Q_2)Q_1 \qquad (3.5)$$

b is the birth rate, a_{11} the 'natural' death rate and a_{12}, the rate of loss through predation. The predator population equation is

$$dQ_2/dt = (-d + a_{21}Q_1)Q_2 \qquad (3.6)$$

where d is the death rate and a_{21} the growth rate determined by the availability of the prey population.

An element of many of these models is a relationship to the logistic growth equation which, in one dimension, is

$$dQ/dt = (K - Q)Q \qquad (3.7)$$

The gradient is zero at $Q = 0$ and $Q = K$ (which are also therefore equilibrium points). This represents logistic growth—an S-shaped curve.

In ecology, K is the 'carrying capacity' of the system for that species and we will be able to interpret terms appropriately on these lines in the models we are developing. It can be seen in the case of Eq. 3.5 that the prey population would grow logistically but for the effects of the predator ($-a_{12}Q_2$) while the predator population has no positive carrying capacity and relies on the prey-food to grow ($+a_{21}Q_1$).

3.4 The Lotka-Volterra, Lanchester, Gause 'Competition for Resources' Model: Ecology and War

Lotka and Volterra also developed a 'competition for resources model. For two species, the equations are

$$dQ_1/dt = (a - a_{11}Q_1 - a_{12}Q_2)Q_1 \tag{3.8}$$

and

$$dQ_2/dt = (b - a_{21}Q_1 - a_{22}Q_2)Q_2 \tag{3.9}$$

In each case, there are constant birth rates but the rate of growth is diminished by the consumption of resources by both species—the different rates represented by the $\{a_{mn}\}$.

The equilibrium conditions are

$$a - a_{11}Q_1 - a_{12}Q_2 = 0 \tag{3.10}$$

and

$$b - a_{21}Q_1 - a_{22}Q_2 = 0 \tag{3.11}$$

These equations are identical to those of Lanchester which are then interpreted to represent models of warfare. a and b represent rates of growth of armies (say), a_{11} and a_{22} limit growth of resources and a_{12} and a_{21} represent attrition from battle. Comparison with the Richardson equations show that these are essentially the same; the difference, however, is that a_{11}, $a_{12} > 0$ in the Richardson model and a_{11}, $a_{22} < 0$ in the Lanchester model—so this change of sign, as noted by Epstein, represents the transition from arms race to war!

These equations can also be written as follows to show, for each population, the negative effects of competition versus logistic growth (with carrying capacities of a/a_{11} and b/a_{22}):

$$dQ_1/dt = -a_{12}Q_1Q_2 + (a - a_{11}Q_1)Q_1 \tag{3.12}$$

and

$$dQ_2/dt = -a_{21}Q_1Q_2 + (b - a_{22}Q_2)Q_2 \tag{3.13}$$

3.5 The Kermack-McKendrick Epidemics 'Threshold' Model

Consider the following very simplified form of model:

$$dQ_1/dt = -a_{12}Q_1Q_2 \qquad (3.14)$$

and

$$dQ_2/dt = a_{21}Q_1Q_2 \qquad (3.15)$$

which can be derived from (3.8) and (3.9) by setting a number of parameters to zero (and changing the sign of a_{21}). This represents the simplest form of epidemics' model: Q_1 is the population of susceptibles and Q_2 the population of infectives. Note that if for a total population, P, we put

$$Q_1 = P - Q_2 \qquad (3.16)$$

then

$$dQ_2/dt = a_{12}Q_2(P - Q_2) \qquad (3.17)$$

showing that the population of infectives grows logistically. In this case, the total population, P is clearly the carrying capacity! However, what usually happens is that infectives are 'removed' from the population—by death, isolation, or recovery, so the model becomes

$$dQ_1/dt = -a_{12}Q_1Q_2 \qquad (3.18)$$

$$dQ_2/dt = a_{12}Q_1Q_2 - r_2Q_2 \qquad (3.19)$$

We can now only have $dQ_2/dt > 0$ if

$$a_{12}Q_1 > r_2 \qquad (3.20)$$

which is

$$Q_1 > r_2/a_{12} \qquad (3.21)$$

This is the Kermack-McKendrick (1927) *threshold* model: the epidemic only takes off if the population of susceptibles is greater than a certain size.

3.6 The Harris-Wilson Model of Urban Development

The model of retail centre evolution described in Chap. 2 was first introduced in Harris and Wilson (1978). This offers two new ideas relative to the earlier models of this chapter. First, it introduces 'space' through the zone systems that we will now be familiar with. Second, it offers the possibility of building a more general model because what would be taken as constant coefficients in the models presented above can now be 'variables' calculated from other submodels. The revenue attracted into a retail centre, D_j, is a striking example of this. Recall that the structural dynamics $\{W_j\}$ model can be presented as

$$\Delta W_j(t,\ t+1) = \varepsilon \big[D_j(t) - C_j(t)\big] W_j(t) \tag{3.22}$$

which, if we take $C_j(t) = KW_j(t)$, is, essentially, a Lotka-Volterra equation. Then

$$D_j = C_j \tag{3.23}$$

at equilibrium. The $\{W_j\}$ in this case can be considered to be either N species or a single species distributed across space.

Note that we could write the model in the form

$$dQ_j/dt = \big[a - a_{11}Q_j\big]Q_j \tag{3.24}$$

for suitable definitions of a and a_{11}. This illustrates the new and powerful point noted earlier: that we can generalise the core equations by making the coefficients *functions*.

3.7 The Bass 'Marketing' Model

The Harris-Wilson model is sometimes presented as

$$dQ_j/dt = \big[a - a_{11}Q_j\big] \tag{3.25}$$

that is, without the Q_j term. Although these two models have the same equilibrium conditions, they can have different trajectories through time because of path dependence (which we will discuss in more detail in later chapters).

The paper by Bass (1969) combines these two elements:

$$dQ/dt = \alpha[a - a_{11}Q] + \beta[a - a_{11}Q]Q \tag{3.26}$$

In his model, dQ/dt is the rate of take-up of a new product; the first term on the right hand side is interpreted as the take up by 'adopters' and the second by 'imitators'—weighted by α and β respectively.

3.8 A General Model

It is always interesting, given the variety of 'different' but 'similar' models that we ask the question: can we write a general model in such a way that each of these are special cases? We now seek to do this with a model with N 'species', n = 1, 2, ...N and space through zones i = 1, 2, ...I, say.

Consider

$$dQ_j^n/dt = \gamma_j^n \left[H_j^n(Q_i^m, u_i^m) - \sum_{im} a_{ij}^{mn} Q_i^m \right] + \varepsilon_j^n \left[H_j^n(Q_i^m, u_i^m) - \sum_{im} b_{ij}^{mn} Q_i^m \right] Q_j^n$$

(3.27)

Or, an even more general model, adding a diffusion term:

$$dQ_j^n/dt = \gamma_j^n \left[H_j^n(Q_i^m, u_i^m) - \sum_{im} a_{ij}^{mn} Q_i^m \right] + \varepsilon_j^n \left[H_j^n(Q_i^m, u_i^m) - \sum_{im} b_{ij}^{mn} Q_i^m \right] Q_j^n$$
$$+ d_j^n \partial^2 Q_{jn}/\partial x^2$$

(3.28)

By appropriate setting of the parameters γ, ε and d, we can generate eight cases:

$\gamma \neq 0$, $\varepsilon = 0$, $d = 0$ (Richardson, Harris-Wilson-1)
$\gamma = 0$, $\varepsilon \neq 0$, $d = 0$ (Lotka-Volterra, Harris-Wilson-2)
$\gamma \neq 0$, $\varepsilon \neq 0$, $d = 0$ (Bass)
$d \neq 0$ some a and b non-zero (reaction–diffusion)
$\gamma \neq 0$, $\varepsilon \neq 0$, $d \neq 0$ (all processes operating)

The first three can be 'with' or 'without' space, hence eight cases.
We now conclude this set of examples by using the general model to identify a number of classic cases into which we can now introduce space.

3.9 A Lotka-Volterra Prey-Predator Model with Space

We now assume the existence of a discrete zone system in the usual way with a set of origin zones, {i} and a set of destination zones, {j}. We add these spatial labels to Eqs. 3.5 and 3.6 in an obvious way. The prey equation is:

$$dQ_j^1/dt = \varepsilon_j^1 \left[K_j^1 - b_{jj}^{11} Q_j^1 - \sum_i b_{ij}^{21} Q_i^2 \right] Q_j^1$$

(3.29)

The prey can now be 'attacked' by the predator from different spatial zones through the term $-\sum_i b_{ij}^{21} Q_i^2$. The coefficient b_{ij}^{21} would therefore have some of the properties of a spatial interaction model. The predator equation is:

$$dQ_j^2/dt = \varepsilon_j^2 \left[- K_j^2 + \sum_i b_{ij}^{12} Q_i^1 \right] Q_j^2$$

(3.30)

A suitable spatial interaction model for the b_{ij}^{21} term might be

$$b_{ij}^{21} = \mu Q_j^1 \exp(-\beta^{21} c_{ik}) \Big/ \sum_k Q_k^1 \exp(-\beta c_{ik}) \qquad (3.31)$$

This shows how the coefficients can become not simply functions, but 'models'.

3.10 An Epstein-Lanchester-Lotka-Volterra 'War' Model with Space

Let us return to the R, B notation for convenience. Consider

$$dR_i/dt = -bR_i \sum_j \lambda_{ij} B_j + \gamma R_i (1 - R_i/K_i) \qquad (3.32)$$

and

$$dB_i/dt = -rB_i \sum_j \mu_{ij} R_j + \varepsilon B_i (1 - B_i/L_i) \qquad (3.33)$$

This adds space to the conventional model by analogy with the prey-predator argument above and so the λ and μ coefficients again have to be spatial interaction models this time representing the 'levels of threat' between countries, the 'reach' of countries at war or the range of armaments in battle.

3.11 Morphogenetic Models

We noted in Sect. 3.8 in Eq. 3.28 that it might be appropriate to add a diffusion term. The immediate provenance of this idea lies in the work of Medda et al. (2009). Medda applied Turing's famous morphogenetic model of 1952, which was designed to represent the evolution of structure in biology, to urban systems. There is a rich literature on spatial diffusion models but the full fruits of Turing's original idea have perhaps not been realized. The essence is this: if there are two inter-acting diffusion processes with different rates, then spatial structures can be cre-ated. The resulting models can be simulated in discrete time though the algorithms are rather tricky because they involve second spatial partial derivatives. An interesting alternative is to seek to integrate the diffusion idea within the spatial interaction-location-structural dynamics framework [using the methods summa-rised, for example, in Wilson (2008)]. We present this in the spirit of showing how to explore a new kind of model development.

We start with a circular employment-monocentric city, as in Medda's work, and then shift to a multi-centric example. Suppose the circle is divided into discrete residential zones, labelled i. Consider the following definitions:

$P_i(t)$ = the population in zone i at time t—and indeed the (t) label can be added to each of the subsequent definitions. For the dynamics, we will consider discrete intervals t, t + Δt, t + 2Δt,.....

$$r_i = r_i(P_i) \tag{3.34}$$

is the rent per unit of population in zone i. It is assumed to be a function of P_i and $\partial r_i/\partial P_i \geq 0$: that is, as the population of a zone increase, so does the rent.

$$c_i = c_i(P_i) \tag{3.35}$$

is the journey to work cost per head of population, also assumed to be a function of P_i, and in this case, it is assumed, following Medda, that if the population increases, there will be investment in transport and the cost will decrease: $\partial c_i/\partial P_i \leq 0$.

$Y_i = Y$ = per capita income in i, assumed constant in the first instance.

$$U_i = U_i(Y - r_i - c_i) = (\text{say})\ (Y - r_i - c_i) \tag{3.36}$$

is the unit utility for a resident of i.

$$U_{ji}(t) = U_{ij}\left[Y - r_j\left(t - \tau_{ji}^R\right) - c_j\left(t - \tau_{ji}^T\right)\right] = \left[Y - r_j\left(t - \tau_{ji}^R\right) - c_j\left(t - \tau_{ji}^T\right)\right] \tag{3.37}$$

is the utility at j as perceived by a resident of i at time t The terms τ_{ji}^R and τ_{ji}^T are introduced to indicate that there will be perceived lags in the perception of $U_{ji}(t)$—R indicating in relation to rent and T in relation to transport—and that these lags will be greater, the greater the distance between i and j. It is assumed that the rent lag will be greater than the transport lag. These lags are the means of introducing diffusion effects since they represent, in effect, the spatial diffusion of information.

d_{ij} is the distance from i to j. The τ's would be functions of d_{ij}. Formally,

$$\tau_{ij}^R = \tau_{ij}^R(d_{ij}) \tag{3.38}$$

$$\tau_{ij}^T = \tau_{ij}^T(d_{ij}) \tag{3.39}$$

$N_{ij}(t,\ t + \Delta t)$ = the number of people who move from i to j in the period (t, t + Δt). $N_{ij} \geq 0$.

A core assumption for the dynamics is then:

$$N_{ji} = \varepsilon\left[U_{ij} - U_j\right] \quad \text{if } U_{ij} - U_j > 0 \tag{3.40A}$$

$$= 0 \quad \text{if } U_{ij} - U_i < 0 \tag{3.40B}$$

That is, a number of people—the scale determined by the parameter ε—will move from j to i if the utility at i, perceived from j, is positive. Similarly,

$$N_{ij} = \varepsilon\left[U_{ji} - U_i\right] \quad \text{if } U_{ji} - U_i > 0 \tag{3.41A}$$

$$= 0 \quad \text{if } U_{ji} - U_i < 0 \tag{3.41B}$$

(3.40A) and (3.41A) can be written explicitly as

$$N_{ji}(t, \ t+\Delta t) = \varepsilon\left[-r_i\left(t - \tau_{ji}^R\right) - c_i\left(t - \tau_{ji}^T\right) + r_j + c_j\right] \tag{3.42}$$

noting that the Ys cancel. Movement will occur if the sum of the perceived rent and travel cost in i is less than the current actual in j.

$$N_{ij}(T, \ t+\Delta t) = \varepsilon\left[-r_j\left(t - \tau_{ij}^R\right) - c_j\left(t - \tau_{ij}^T\right) + r_i + c_i\right] \tag{3.43}$$

Then

$$\Delta P_i(t, \ t+\Delta t) = \Sigma_j\left[N_{ji}(t+\Delta t) - N_{ij}(t+\Delta t)\right] \tag{3.44}$$

$$P_i(t+\Delta t) = P_i(t) + \Delta P_i(t, \ t+\Delta t) \tag{3.45}$$

We need to specify the functional forms of r_i and c_i in (3.34) and (3.35) and τ_{ij}^R and τ_{ij}^T in (3.38) and (3.39). Then the model can in principle be solved for any set of initial conditions by cycling from (3.33) to (3.45) and then re-cycling. It would obviously be possible to seek to derive an equivalent of Medda's results by starting with all the r_i and c_i equal and the exogenously reducing one of the c_i. In the first instance, the r_i and c_i can be any plausible functions. The τ's need to increase with distance and might be taken as $\kappa^R\exp(\gamma^R d_{ij})$ and $\kappa^T\exp(\gamma^T d_{ij})$. It should then be possible to run simulations of the (3.33)–(3.45) model.

The τ-terms are the basis for any 'diffusion' interpretation. The N_{ij} are 'currents' and $N_{ji} - N_{ij}$ is a difference in currents—but not a second spatial derivative of the same current. On the other hand if $\Delta P_i(t, \ t+\Delta t)$ in (3.43) is written out in full using Eqs. 3.41 and 3.42, we have something like

$$\Delta P_i(t, \ t+\Delta t) = \varepsilon\left[-r_i\left(t - \tau_{ji}^R\right) - c_i\left(t - \tau_{ji}^T\right) + r_j + \ c_j + r_j\left(t - \tau_{ij}^R\right)\right.$$
$$\left. + cj\left(t - \tau_{ij}^T\right) - r_i - c_i\right] \tag{3.46}$$

in which the right hand side could, perhaps look like, a second spatial derivative in a discrete space formulation—but this is speculation and needs further investigation.

Particular care should be taken in both the simulation and the exploration of any interpretations, of Eqs. 3.40 and 3.41 because this introduces discontinuities in the definitions of the elements of the $\{N_{ij}\}$ array—arising from the need to keep element greater than or equal to zero.

A further area for future investigation is to build on the introduction of the utility function, to make it explicit also, and to carry it into a full economic analysis that includes the introduction of demand functions.

We can now proceed to the two-dimensional case. The notation for the 1-D case above has been set up in such a way that it extends easily. Most of it still holds good. We could retail the assumption of a single employment centre. $\{d_{ij}\}$ would be a 'richer' matrix—as then would $\{\tau_{ij}^R\}$ and $\{\tau_{ij}^T\}$—but the model would still run as in Eqs. 3.33–3.45. To move to a fully multi-centric model, it would be necessary to specify employment by zone and a journey to work array, say $\{T_{ij}\}$. This could be taken as data, but it would almost certainly be better to take it as specified by a model:

$$T_{ij} = A_i B_j P_i E_j \exp(-\beta c_{ij}) \tag{3.47}$$

with

$$A_i = \sum_j B_j E_j \exp(-\beta c_{ij}) \tag{3.48}$$

and

$$B_j = \sum_i A_i P_i \exp(-\beta c_{ij}) \tag{3.49}$$

to ensure

$$\Sigma_j T_{ij} = P_i \tag{3.50}$$

and

$$\Sigma_i T_{ij} = E_j \tag{3.51}$$

in the usual way. $\{E_j\}$ would be a given distribution of employment.

There are then two ways of proceeding. We could calculate an average journey to work cost in i from

$$c_i = \Sigma_j T_{ij} c_{ij} / \Sigma_j T_{ij} \tag{3.52}$$

and then proceed as with Eqs. (3.33)–(3.45) using this c_i and the new d- and τ-matrices. Or we could develop a fuller more explicit model by taking i as the residential zone, k as the employment zone, and considering (i, k) to (j, m) transitions. The core of the model would then be

$$\Delta P_{ik} = \Sigma_{jm} \left[N_{jmik} - N_{ikjm} \right] \tag{3.53}$$

for a suitably extended $\{N_{ikjm}\}$ array. This would imply utility definitions that enabled N_{ikjm} to be calculated from

$$N_{ikjm} = \varepsilon\left[U_{jmik} - U_{ik}\right] \qquad (3.54)$$

References

Bass FM (1969) A new product growth model for consumer durables. Manage Sci 15:215–227

Gause GF (1934) The struggle for existence. Williams and Wilkins, Baltimore

Harris B, Wilson AG (1978) Equilibrium values and dynamics of attractiveness terms in production-constrained spatial-interaction models. Environ Planning A 10:371–388

Kermack WO, McKendrick AG (1927) A contribution to the mathematical theory of epidemics, Proceedings Royal Society of London A 115:700–721

Lanchester FW (1916) Aircraft in warfare: the dawn of the fourth arm. Constable, London

Lotka AJ (1924) Elements of physical biology. Williams and Wilkins, Baltimore

Richardson LF (1939) Generalized foreign politics, The British Journal of Psychology, Monograph Supplement Number 23

Richardson LF (1960) Arms and insecurity: a mathematical study of the causes and origins of war. Boxwood Press, Pittsburgh

Turing A (1952) The chemical basis of morphogenetics. Philos Trans R Soc 237:37–54

Volterra V (1938) Population growth, equilibria and extinction under specified breeding conditions: a development and extension of the theory of the logistic curve. Hum Biol 10:1–11

Further Reading

Epstein JM (1997) Nonlinear dynamics, mathematical biology, and social science. Addison-Wesley, Reading

Medda F, Nijkamp P, Reitveld P (2009) A morphogenetic perspective on spatial complexity: transport costs and urban shapes. In: Reggiani A, Nijkamp P (eds) Complexity and spatial networks. Springer, Berlin, pp 51–60

Wilson AG (2006) Ecological and urban systems models: some explorations of similarities in the context of complexity theory. Environ Planning A 38:633–646

Wilson AG (2008) Boltzmann, Lotka and Volterra and spatial structural evolution: an integrated methodology for some dynamical systems, J R Soc Interface 5:865–871. doi:10.1098/rsif.2007.1288

Chapter 4
Theory

4.1 Introduction

Models are representations of theories. Theories are built out of concepts and a knowledge of these concepts should be part of our tool kit. At the lowest level, these concepts are essentially system descriptors—and many of our model variables fall into this category; at higher levels, they can function as elements of theories which give us a deeper level of understanding. As a framework, we can take Fig. 1.1 from Chap. 1 which sets out the main elements of an urban or regional system and which is repeated here for convenience (Fig. 4.1).

The functioning and evolution of the system has three roots: the population and their activities, levels of economic activity, and the infrastructure that carries these activities. Demography, economics and geography will all function as important disciplines in assembling a tool kit of concepts but, as we will see, there are other significant contributors. A first approach to the building blocks of theories, therefore, is to examine the ways in which different disciplines have contributed and we do this in Sect. 4.2. We can then use this survey as the basis for an interdisciplinary assessment of what is needed for theory building, which we do in Sect. 4.3.

4.2 Core Concepts by Discipline

4.2.1 Demography

We have already noted the importance of accounts and counting the populations of our systems of interest and the ways in which these counts are changing is an obvious starting point. The mechanisms of change are through birth, death and migration. The last of these should perhaps be more properly labelled

A. Wilson, *The Science of Cities and Regions*, SpringerBriefs in Geography,
DOI: 10.1007/978-94-007-2266-8_4, © The Author(s) 2012

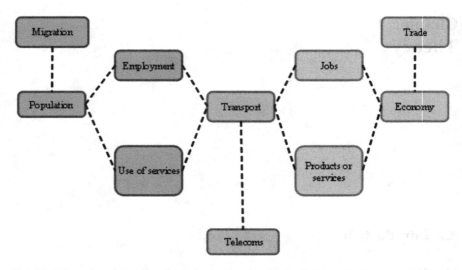

Fig. 4.1 The main subsystems of an urban and regional system

'change of location' since we normally retain 'migration' to refer to international movements while intra-national or intra-regional residential relocation is potentially very important. As ever, our treatment of core concepts depends on scale. At a global scale, the focus will be on inter-country migration; at an urban scale, or residential relocation—for example, the movement first out of, and then back to, inner cities.

Demographic accounts are relatively straightforward and we illustrate the basics in Chap. 5. What is much more difficult is forecasting the change in birth, death and migration rates over time. These are essential inputs to the population components of models. There are further complications in that for many purposes, we need an understanding of household structures and how they evolve. So the underpinning theory we are seeking relates to these core rates. There is a huge amount of statistical data—from population censuses for example—on births and deaths and, to a lesser extent, migration. Resulting statistical analyses show many correlations: higher birth rates are associated with lower incomes, but so also are lower-age death rates. This is a case where the statistical models probably produce sufficiently robust predictions to serve as the inputs to the mathematical models which constitute our subject matter.

The case of migration is different in that spatial interaction models come into play. The theory seeks to identify 'push' and 'pull' factors that drive migration and distance or cost, coupled with different kinds of cultural histories and ties between countries also play a role. These are all factors that can easily be incorporated into models and this has indeed been done.

4.2.2 Economics

Underpinning economic concepts are crucial to effective modelling of cities and regions. Traditional theory in economics, and the associated concepts, subdivides according to scale: the micro relating to individuals or households or to organisations; the macro, to regional or national economies. In the micro case, we need concepts that will underpin theories of how people, households and organisations behave in an economic environment. Typically, individuals are assumed to maximise utility; and firms, in a market system, adjust inputs to their production functions, in the light of demand, to maximise profit. It is more difficult to characterise the economic objectives of public service organisations though attempts are made through the estimation of social benefits (in financial terms) within the framework of cost-benefit analysis. This then puts the emphasis on defining and estimating utility functions, production functions and social benefits. The macro scale is more likely to be connected to broader 'sectors' of an economy and how they are related and with measures of economic activity such as gross domestic product (GDP). There is a fundamental difficulty in linking the micro to the macro—the 'aggregation problem in economics'—cf. Green (1964) and Arrow (1951).

We must then note that much of what interests us in urban and regional modelling is what might be called the meso scale. This arises from our use of spatial systems to enable us to identify interaction and evolving spatial structures. We can then choose, in different circumstances, whether to locate micro units in these arrangements or macro sectors. We have presented this as a model design issue earlier: choosing an appropriate level of disaggregation. The 'aggregation problem' remains with us! Economists have been less effective at this scale. It can be argued that this has arisen from poor choice of spatial system—a tendency to use continuous rather than discrete spatial representations—and also rather artificial representations of systems—too many people for example assumed to be behaving in some 'economic rational' way when clearly they don't. This is not because people are 'irrational', but because if their choices arise from utility maximisation, their utility functions are much more varied than conventional economic theory can represent.

A useful perspective on what modellers can extract from economic theory is provided by Paelink and Nijkamp (1975), quoted in Wilson (2000). They argue that what economics can contribute to urban and regional analysis is based on seven principles—paraphrased here, and presented in italics. Short commentaries are added in each case on how we deploy these principles (or, in one case, not!).

1. *A focus on consumption and production processes—in effect, seeking to specify production and utility functions subject to any constraints—for example in relation to available technologies.* We will see later that, for example, the $W_j^\alpha \exp(-\beta c_{ij})$ term in the retail model can be related to the idea of a utility function, and the KW_j term in the dynamic retail model to a production

function. This shows how we can strengthen our models by relating our concepts to economic ones.

2. *Charting substitution possibilities—different combinations of inputs that produce the same outputs; hence the emphasis in much economic modelling on elasticities of substitution.* This is not an extensively applied principle in modelling and so suggests an area for further research. The nearest representation of this kind of substitution is probably in modal choice in transport models.

3. *Optimising principles: maximising profit, minimising cost, maximising utility, maximising social benefits.* These are implicit in many urban and regional models and have an interesting relation to entropy-maximising. See the comment in (4) below and in Chap. 6.

4. *Recognising that markets are imperfect.* In the argument of these lectures, 'entropy' plays a critical role here because this is what makes the optimising principle in spatial interaction models represent imperfect markets. Again, see Chap. 6 below.

5. *Recognising economies of scale—agglomeration economies—arising, for example from indivisibilities in production processes.* This is partly recognised in modelling through the use of 'constraints' which are both a valuable weapon, but also cause difficulties in handling, in model building. Positive returns to scale are shown, for example in the $\alpha > 1$ case when W_j^α is used as an attractiveness factor in spatial interaction models.

6. *Recognising external economies—e.g. the impacts of pollution generated by one producer on other producers.* This is not well recognised or represented in modelling.

7. *The importance of transport costs.* These are a critical element of spatial interaction models.

While, as indicated, progress has been made, it remains a major research challenge to integrate the concepts of economics with the bulk of urban and regional modelling.

4.2.3 Geography

Geography potentially fills the major gap left by economics in that it focuses on space and in this sense can be considered to take the meso scale seriously. The focus is on location of activities, interaction, networks and settlement structures for example—all ranging from neighbourhood to global scales. However, geography is still concerned with the same subsystems as economists and so all the associated concepts need to be brought to bear.

Particular concepts relate to the settlement size distributions, market areas and catchment populations for example. Each of the subsystems of a comprehensive urban and regional system has a long modelling history. Each has aspects of the

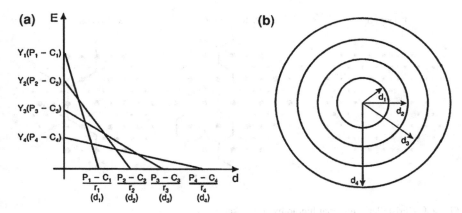

Fig. 4.2 von Thunen's rings

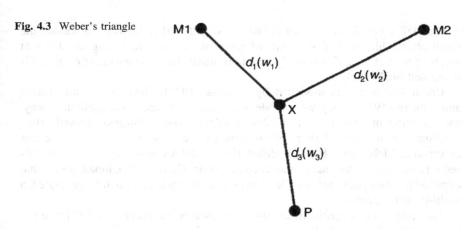

Fig. 4.3 Weber's triangle

two key features of our models: location and interaction. von Thunen (1826), in modelling agricultural land use, focused on the interaction between farmer and market and then predicted locational output: crop by zone. He introduced the concept of 'bid rent' which we will find playing a role in Chap. 6. It was developed by the urban economist William Alonso (1960) and then transformed into a mathematical model with discrete zones by Herbert and Stevens (1960). von Thunen had one market centre in his model—'the isolated state'—and he predicted rings of different agricultural land uses as shown in Fig. 4.2. The left hand part of the Figure shows the bid rent lines against distance from the market for different kinds of crops.

Weber (1909), in seeking to model industrial location, had a double interaction: inputs to factories and outputs from factories to markets. His locational output was the product mix at factories and, in particular, their location. At its simplest, this generated the famous Weber triangle shown in Fig. 4.3.

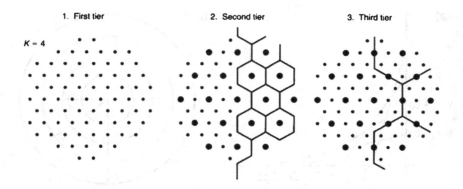

Fig. 4.4 Christaller's regional hierarchy of market areas

Christaller (1933) and Losch (1940), in slightly different ways, modelled the mutli-centric hierarchical structure of regional systems, focussing on different levels of market area. Christaller's system in particular was elegant, if rigid and is illustrated in Fig. 4.4.

Urban structure was modelled by Burgess (1927), Hoyt (1939) and Harris and Ullman (1945). They were modelling social structures in a qualitative way, each offering insights into the modes of urban growth. Burgess showed urban development in terms of rings of invasion and succession—do we capture this in urban models now? Hoyt modified this model by showing that sector differentiation had to be taken into account while Harris and Ullman noted that expanding cities absorbed smaller towns and villages and so this generated a multi-centric structure.

The early work on spatial interaction was based on the gravity model. There is a long history—Carey (1858) and Ravenstain (1885)—who applied it to migration, Lill (1891) and Young (1924). Reilly (1931) applied the gravity idea to retail and so was the precursor of our Chap. 2 model. The models were formalised by Stewart (1942) and Zipf (1946) who connected the model to retailing, through market areas. However, it was not until the mid 1960s that it was recognised that this also functioned as a location model as we have seen.

Each of these 'classical' models can be reformulated in the X-Y-Z notation of Chap. 2 and generalised. The restrictive assumptions, such as the single market centre in von Thunen's model or the rigid geometry of non-overlapping market areas in Christaller's can then be relaxed. The detailed argument is set out in Wilson (2000).

4.2.4 *Other Social Sciences*

We group together sociology, politics and psychology (and perhaps anthropology) which all have related subject matter. They are concerned with individual behaviour within social, cultural and political environments and the study of these environments through such concepts as 'social movements'. If individuals are thought of as 'agents', then at a micro scale, they function within (exogenously) given structures. This leads to the idea of the 'agency-structure' problem: how structures are created by, or at least influenced by, individual behaviour. This is the sociological analogue of the aggregation problem in economics.

The focus on individual behaviour demands a categorisation of different kinds of people and households—for example in the notion of 'social class'. Since these categorisations are a crucial part of model building, this is territory where we might look for help.

4.2.5 *History*

Giddens (1979)—again quoted in Wilson (2000)—argued that there is no distinction between history and sociology. The subject matter is essentially the same. This raises interesting question about the development of theory in history. However, from model design and building perspectives, the importance of the historian is in assembling data on urban and regional evolution over long time periods. This provides the basis for developing dynamic models. It is also the case, of course, that the good historian will offer hypotheses to explain this evolution and this should provide further insights for the modeller. An excellent example on both counts is provided by Cronon's (1992) book *Nature's metropolis*. This was the basis of the model example offered at the end of Chap. 2 on the evolution of the North American urban system.

4.2.6 *The Physical Environment*

Physical environments have not played much role hitherto in the urban and regional modelling world beyond building in topography to network representation. This should now begin to change as so may of the big strategic issues are rooted in the environment. Climate change is an obvious example. It is relatively straightforward, for example, to estimate pollution levels within models and to test alternative scenarios—e.g. in transport and urban densities—in relation to carbon emissions for instance. There are other issues, such as competition for resources such as water and oil; and food security.

We have also indicated that we can pick up modelling ideas from some of these fields—most notably ecological modelling—and vice versa: we have something to offer these fields.

4.2.7 Mathematics and Statistics

These disciplines are fundamental to the tool kit. For example, because so many, if not all of the models, have nonlinearities, including positive returns to scale, it is important to identify the key generic features of such systems: multiple equilibria', dependence on initial conditions, and hence path dependence; and phase changes. However, we pursue these issues in more depth in Chap. 5.

4.3 Interdisciplinary Theory Building

Urban and regional models have foundations that are essentially interdisciplinary. We need to have an understanding of available theory—from whatever disciplinary or interdisciplinary source—to achieve, for example, the most effective classification and categorisation of the model's elements; hypotheses that are appropriate to the model's scale; hypotheses that are maximally compatible with known underpinning theory (bearing in mind that there is often no consensus on the theory). From an economic perspective, since most of our models are not in the traditional sense 'economic', we need to find ways of bundling the elements of utility into a small number of groups and to relate these to such categories as income. We need to represent employment, housing and access to other facilities and hence to be connected to labour market theory, housing market theory and public sector economics.

All of this should be related to the workings of our system of interest. In terms of system evolution, we should pay particular attention to charting chains of causality—perhaps even formally using such techniques as graphical models. Above all, we need to understand the drivers of change—and this needs an expansion of the Paelink-Nijkamp list. Consider some of the obvious drivers that can impact on model development: migration and relocation; income growth; car ownership; technological change; climate change; availability of resources: water, energy; security issues.

References

Alonso W (1960) A theory of the urban land market. Pap Reg Sci Assoc 6:149–157
Arrow KJ (1951) Social choice and individual values. Wiley, New York
Burgess EW (1927) The determinants of gradients in the growth of a city. Publ Am Sociological Soc 21:178–184

Carey HC (1858) Principles of social science. Lippincott, Philadelphia

Christaller W (1933) Die centralen Orte in Suddeutschland, Gustav Fischer, Jena; translated by Baskin CW Central places in Southern Germany, Prentice Hall, Englewood Cliffs

Cronon W (1992) Nature's metropolis: Chicago and the great west. W.W. Norton, New York

Giddens A (1979) Central problems in social theory. University of California Press, Berkeley

Green HAJ (1964) Aggregation in economic analysis. Princeton University Press, Princeton

Harris CD, Ullman EL (1945) The nature of cities. Ann Am Acad Political Social Sci 242:7–17

Herbert DJ, Stevens BH (1960) A model for the distribution of residential activity in an urban area. J Regional Sci 2:21–36

Hoyt H (1939) The structure and growth of residential neighbourhoods in American cities. Federal Housing Administration, Washington

Lill E (1891) Das reisegesetz und seine anwendung auf den eisenbahnverkehr, Wien; cited in Erlander and Stewart (1990)

Losch A (1940) *Die raumliche ordnung der wirtschaft*, Gustav Fischer, Jena; translated by Woglam WH, Stolper WF (1954) The economics of location, Yale University Press, New Haven

Paelink JHP, Nijkamp P (1975) Operational theory and method in regional economics. Saxon House, Farnborough

Ravenstein EG (1885) The laws of migration. J Roy Stat Soc 48:167–227

Reilly WJ (1931) The law of retail gravitation. G.P. Putman, New York

Stewart JQ (1942) A measure of the influence of population at a distance. Sociometry 5:63–71

von Thunen JH (1826) Der isolierte staat in beziehung auf landwirtschaft und nationalokonomie, Gustav Fisher, Stuttgart; English translation. In: Wartenburg CM (ed) The isolated state. Oxford University Press, Oxford (1966)

Weber A (1909) Uber den standort der industrien, Tubingen; English translation. In: Friedrich CJ (ed) Theory of the location of industries. University of Chicago Press, Chicago

Young EC (1924) The movement of farm population, bulletin 426, agriculture experiment station. Cornell University, Ithaca

Zipf GK (1946) The P_1P_2/D hypothesis on the inter-city movement of persons. Am Sociological Rev 11:677–686

Further Reading

Wilson AG (2000) Complex spatial systems. Prentice Hall, Englewood Cliffs, Chapter 7

Wilson AG (2010) Knowledge power. Routledge, London

Chapter 5
Methods: The Model-Building Tool Kit

5.1 Introduction

Probably as with all tool kits, there are different ways of assembling the elements, with different mixes, to tackle a particular task. We choose here to present the tools in three boxes. First we focus on account-based models. So many models are rooted in accounts—literally counting the elements of our systems of interest and how these numbers change—that this makes a good starting point. We have seen from the examples so far that we will usually use a discrete zone system with variables like P_i representing the population of a typical zone i and $\{T_{ij}\}$ or $\{S_{ij}\}$ representing interaction between zones. We will seek to draw on this kind of notation to illustrate the accounts as they are introduced. Secondly, we review the mathematical tools that are available to us. We note briefly a range of modelling styles which can be described as 'generic'. These have often been applied in one field but can be applied more widely. We then, thirdly, review the basic mathematical methods that are available to us—essentially as an outline of a mathematics-for-modellers 'catch-up' course!

5.2 Account-Based Frameworks

5.2.1 Introduction

As we have noted, many of our models are rooted in accounts and it is helpful to make this explicit because such accounts are often the foundations of model design. In some instances for particular systems of interest, the models are essentially generic for a wider class of situations and so they function as elements of a basic tool kit. We discuss in turn demographic models; economic input–output models; spatial interaction models and location models.

A. Wilson, *The Science of Cities and Regions*, SpringerBriefs in Geography, DOI: 10.1007/978-94-007-2266-8_5, © The Author(s) 2012

5.2.2 Demographic Models

Demographic accounts are obviously useful in population analysis and forecasting but potentially can be used for any system involving birth, death and migration— ecological systems for example. The core form of the multi-regional model was developed by Rogers in the 1970s, building on the cohort-survival model of Leslie in the 1940s. In their most elaborate form, they were presented in Rees and Wilson (1977) and many applications have been developed on these premises.

In demographic modelling, it has to be recognised that the different rates will be age-dependent and it therefore becomes important to represent age-cohorts explicitly. To illustrate the modelling principle, we follow Wilson (1974, pp. 80–82), but make the notation more consistent with that used in the rest of this book. Suppose we label age groups by r and s and initially, we ignore space and let $P^r(t)$ be the population of some system in group r at time t. Let $b_r(t, t + 1)$ be the birth rate for group r which will be non-zero between $r = a$ and $r = b$, say, the limits of the child bearing age groups; let s^{r-1r} be the rate of survival from $r - 1$ to r in t to $t + 1$ and m^{r-1r} be the net immigration rate in the period. Then

$$P^1(t+1) = \sum_a^b b^k(t, \ t+1)P^k(t) \tag{5.1}$$

for the first age group and

$$P^r(t+1) = s^{r-1r}(t, \ t+1)P^{r-1}(t) + m^{r-1r}(t, \ t+1)P^{r-1}(t) \tag{5.2}$$

for $r > 1$.

It is then relatively straightforward to add space. We add zone labels as subscripts in the usual way and then Eqs. 5.1 and 5.2 become

$$P_i^1(t+1) = \sum_a^b b_i^k(t, \ t+1)p_i^r(t) \tag{5.3}$$

and

$$P_i^r(t+1) = s_i^{r-1r}(t, \ t+1)P_i^{r-1}(t) + \sum_{j \neq i} m_{ij}^{r-1r}(t, \ t+1)P_j^{r-1}(t) \tag{5.4}$$

These equations represent the basic accounts. It is immediately clear that the rates are core components and need themselves to be modelled, as noted earlier in Chap. 4. These models represented by (5.1)–(5.4), as will be clear to those familiar with matrix algebra, can be effectively represented in matrix form but that takes us beyond the scope of this chapter. We should note, however, that when this is done, the $r - 1$ and r superscripts are reversed to facilitate appropriate matrix multiplication.

What we can do now is to show the accounts' structure that should underpin any demographic model. These principles are fully articulated in Rees and Wilson (1977). This is an interesting example because it demonstrates the implications of

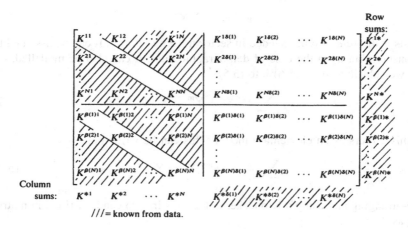

Fig. 5.1 Fully-comprehensive spatial demographic accounts. (From Spatial Population Analysis p. 41)

seeking to be fully explicit in accounts. This arises because there are potentially more 'events' in a time period than we have accounted for in the (5.1)–(5.4) equations: a person born in zone i may migrate to another zone, j, during the period; a person may migrate from i to j but not survive there; a person may be born in the period but not survive until the end of the period (and may even migrate in between). These events are shown, using an obvious notation, in a set of accounts in Fig. 5.1. An i or a j means either existed at the beginning of the period or survived to the end of the period; ij, i ≠ j, implies a migration; β(i)j means 'born in i during the period and survived in j at the end of the period; iδ(j) means 'existed in i at the start of the period but died in j during the period'; β(i)δ(j) means 'born in i during the period but died in j before the end of the period'.

This kind of accounting framework could be used in other areas—for example to model social mobility. It would be necessary to define social class and/or income groups and then to model transition between states.

5.2.3 Economic Input–Output Models

Economic accounts at the aggregate, sector, level are very important. One way or another they underpin any good urban and regional model. The core elements, for a single region, are X^m, the total product in sector m, Y^m, the final demand for that product, and an array, Z^{mn}, the flow of intermediate goods or services from sector m to sector n. The basic accounts are then

$$\sum_n Z^{mn} + Y^m = X^m \tag{5.5}$$

and a rudimentary model can be created by assuming constant rates

$$a^{mn} = Z^{mn}/X^n \qquad (5.6)$$

This is the *rate* of input needed in sector n from sector m. It can be assumed to be constant, but as in the case of demographic rates, it can also be modelled.

If we substitute from (5.6) into (5.5),

$$\sum_n a^{mn}X^n + Y^m = X^m \qquad (5.7)$$

and this servers as an elementary model. This can be written as

$$\sum_n [\delta_{mn} - a^{mn}]X^n = Y^m \qquad (5.8)$$

by re-arranging and using a Kronecker delta. This in turn can be written in matrix form as

$$X = [I - A]^{-1}Y \qquad (5.9)$$

The multi-regional model can be built by turning each of the Zs into matrices as we have already indicated in building a system description in Chap. 1. Note that there are applications at different scales: most usually country, but also 'regional' and urban.

Alternative assumptions can be made about the drivers of the model—the 'rates'—and this illustrates the 'chains of causality' argument. When spatial interaction models are introduced for trade, a more complex model has to be built, ideally from an entropy-maximising framework with the usual constraints for the flows but with the addition of the input–output accounting equations also as constraints.

5.2.4 A Family of Spatial Interaction Models

Spatial interaction models are account-based through their row and column sums. Consider the matrix $\{T_{ij}\}$ in Fig. 5.2.

Here we show the sum of the ith row to be O_i—the total number of trip origins—and the sum of the elements of the jth column to be D_j—the total number of trip destinations. There is a key decision to be made in the design of spatial interaction models in relation to these accounts: do we know one or both of the vectors $\{O_i\}$ and $\{D_j\}$—or indeed neither—independently of the spatial interaction model. We could say, for example, that they were both known from data and that we were building a model to estimate the flow matrix given this data. Or one or both could be separately estimated from a sub model. There are clearly four cases:

1. Both are known, in which case the interaction model is said to be doubly constrained;
2. $\{O_i\}$ is known, and the model is production constrained;

Fig. 5.2 A spatial interaction
matrix showing row and
column sums

T_{11} T_{12} T_{13} .... O_1

T_{21} T_{22} T_{23} .... O_2

....

....

T_{N1} T_{N2} T_{N3} .... O_N

D_1 D_2 D_3 .... D_N

3. $\{D_j\}$ is known and the model is attraction constrained;
4. neither is known, so the model is unconstrained.

Note that we have used the term 'production' instead of 'origin' and 'attraction' instead of 'destination', though this is a matter of taste and historical usage. Cases (2) and (3) are sometimes described, generically, as 'singly constrained'.

This brief presentation also illustrates another general model building point to which we return from time to time. That is, constraints represent some of our (semi) independent knowledge of the system of interest and this knowledge should be built into the model design. This is particularly important for spatial interaction models as we will see in the in-depth discussion in Chap. 6.

5.2.5 Location Models

In the singly constrained spatial interaction model cases, as we have already seen with the retail model, the interaction model also functions as a location model. In the production constrained case, for example, D_j can be calculated as

$$D_j = \sum_i T_{ij} \qquad (5.10)$$

These ideas could be represented in the X-Y-Z-W representation, of course, with the X's and Z's as row and column sums of the Y's. We have already noted a production constrained model functioning as a retail model. There are other applications in relation to residential location—which might be attraction constrained—all the 'workers' have to be located in residences somewhere. And there are applications in terms of the location of public facilities, where the interaction model might be constrained, as in the retail case, at the residential end.

5.3 Mathematical Tools

5.3.1 Generic Approaches

In assembling appropriate mathematical tools, it is helpful to start by noting that certain types of system lend themselves to particular approaches to modelling. We need to bear in mind Weaver's classification of problem/system types. Recall these categories are: simple, of disorganised complexity and of organised complexity. Simple systems can be handled with the basic mathematics of algebra and calculus. Systems of disorganised complexity can be handled through statistical averaging methods—essentially those of Boltzmann's statistical mechanics. Systems of organised complexity involve nonlinearities and, as we have already seen, this takes us into more demanding territory.

Boltzmann's methods generally apply to systems of disorganised complexity—large numbers of elements, weakly interacting, so that statistical averaging methods can be used. We explore these methods in detail in Chap. 6.

Ecological, environmental and epidemiological systems are usually systems of organised complexity and demand the methods on nonlinear mathematics. We saw examples of prey-predator models, competition-for-resources models, warfare models, retail models and epidemiological models in Chap. 3, each with some kind of relationship to the Lotka–Volterra equations.

We now proceed by briefly reviewing the basic mathematics that should be in the modellers' tool kit and then examine various specialist techniques that have roles in different kinds of modelling. We discuss in turn mathematical programming, network analysis, rule-based representations, nonlinear dynamics and various combinations. What follows is essentially (and necessarily) an outline for a focused maths course! What this demonstrates is that it is possible to assemble a mathematics tool kit without taking on board the whole apparatus of a mathematics degree!

5.3.2 Basics

We should take for granted the basics of algebra, geometry and the theory of functions. Our examples and notation already take us into the realms of matrix algebra. A next step is the calculus, differential and integral—and especially in relation to differential and difference equations. These enable us to represent the basics of exponential growth and logistic growth for example and are crucial to our understanding of dynamic models. With time and space both usually represented as discrete, this takes us into the area of finite difference methods for dynamic modelling. We look to differential calculus of course for the mathematics of dynamical systems, and in particular, nonlinear systems. We have already seen from the example in Chap. 3 that these systems exhibit multiple equilibria, path dependence and phase changes.

The calculus is also the basis of optimisation and many of our models can be presented as maximisation or minimisation problems. We take this further in the discussion of mathematical programming below.

The basics of statistics are important to us, both in understanding data and in calibrating models. These basics range from regression analysis and models through to cluster analysis and discriminant analysis as the basis of geodemographics. This also connects to the 'chains of causality' argument and the art of building model equations through functional relationships. There are special cases, such as Markov models–statistical models in which the state at time $t + 1$ is dependent only on the state at time t—and a number of account-based models can be represented in this form. The Bayesian statistics perspective is important for us in that it has a close connection to entropy-maximising models and this in turn links to graphical models and Bayesian networks and, again, as another approach to the 'chains of causality' questions.

5.3.3 Mathematical Programming

Any model that can be represented as an optimisation problem—maximisation or minimisation—falls in the category of mathematical programming and there is a considerable body of work available for the tool kit. There are a number of classical models. One of the best known is the transportation problem—in this case a linear programming example: given a set of origin totals, $\{O_i\}$ say and destination totals, $\{D_j\}$ and transport costs $\{c_{ij}\}$ what is the set of flows that minimise total transport costs $\sum_{ij} T_{ij} c_{ij}$. Formally, the problem is:

$$\text{Min } \{T_{ij}\} \ C = \sum_{ij} T_{ij} c_{ij} \tag{5.11}$$

such that

$$\sum_j T_{ij} = O_i \tag{5.12}$$

and

$$\sum_i T_{ij} = D_j \tag{5.13}$$

We develop this further in Chap. 6.

These kinds of mathematical programming problems have been a mainstay of operations research. The transportation problem, for example, has obvious applications for freight companies. Problems can be formulated for facility location and again, we return to such cases in Chap. 6. It is possible of course, indeed likely as we will see, that either the objective function or the constraints (or both) will include nonlinear functions. Again, there is a considerable mathematical tool kit available to support model development. In the case of linear programmes, it is not possible to write down solutions analytically, but computer algorithms are available to solve many of the problems. In the nonlinear case, analytical solutions can sometimes be found.

One of the most remarkable features of mathematical programming is that each primary problem has a dual: the objective function becomes a constraint; and each constraint becomes a variable in the dual objective function. In the case of the transportation problem, for example, the dual of (5.11)–(5.13) becomes

$$\text{Max } \{\alpha_i, v_j\} \ C' = \sum_i \alpha_i O_i + \sum_j v_j D_j \qquad (5.14)$$

subject to

$$c_{ij} - \alpha_i - v_j \geq 0 \qquad (5.15)$$

The dual variables can be interpreted as 'rents' or 'comparative advantage' and this will help in providing an economic interpretation of models that can be formulated in mathematical programming form.

5.3.4 Network Analysis

As we have seen, many of our models involve spatial interaction—flows between origins and destinations. These flows are carried on networks and so these become integral parts of our models. We should make a distinction at the outset between real networks, such as road or public transport networks and notional networks—the latter representing, for example, notional links between nodes. The two come together with the idea of 'spider networks' which are used as representations of real networks by connecting nearby nodes with notional links.

Our models turn on some measure of travel cost between zones. Ideally, these should be calculated as the sum of such costs along the best route. There are well-known algorithms for solving this problem. We should also bear in mind that in practice, there will be dispersion across 'near best' routes. Once routes have been established, travel costs calculated and the spatial interaction model run to calculate flows, the flows can be assigned to the network. Note that for any particular *link*, there will be typically many routes using that link, and so link flow is a sum of origin–destination flows. This is important if there are congestion elements in the travel costs and, indeed, potentially leads to iteration within the models: run the interaction model, assign the flows, recalculate the costs to take account of congestion and then rerun the model. To approach this formally, consider a set of zones, {i} with these elements can also standing as names for zone centroids. We will also use j, u and v as centroid labels. For example, (u, v) may be a link on a route from i to j on the spider network. (u, v) ε R_{ij}^{min} is the set of links that make up the best route from i to j. This may involve a mix of links of different levels.

Let γ_{uv} be the 'cost' of travelling on link (u, v). Let c_{ij} be the generalised cost of travel from i to j as perceived by consumers. So

$$c_{ij} = \sum_{(u,v)\varepsilon \,\mathrm{Rijmin}} \gamma_{uv} \qquad (5.16)$$

If O_i and D_j are the total number of origins in zone i and destinations in zone j, and T_{ij} is the flow between i and j, then the standard doubly constrained spatial interaction model is

$$T_{ij} = A_i B_j O_i D_j \exp(-\beta c_{ij}) \qquad (5.17)$$

with

$$A_i = 1/\sum_k B_k D_k \exp(-\beta c_{ik}) \qquad (5.18)$$

$$B_j = 1/\sum_k A_k O_k \exp(-\beta c_{ki}) \qquad (5.19)$$

To model congestion, we need to know the flows on each link. Let q_{uv}^h be the flow on link (u, v, h) and let Q_{uv}^h be the set of origin–destination pairs at level h that use the (u, v, h) link. Then

$$q_{uv}^h = \sum_{ij\varepsilon Quvh} T_{ij} \qquad (5.20)$$

We then want γ_{uv}^h to be a function of the flow:

$$\gamma_{uv}^h = \gamma_{uv}^h(q_{uv}^h) \qquad (5.21)$$

These equations then have to be solved iteratively. There is a related and substantial challenge: how to model the evolution of networks and we return to this in Chap. 9.

We can use this example of the core spatial interaction model to describe an approach to presenting the topology of a network. There are now many such approaches but a very useful and simple one is that of Nystuen and Dacey (1961). From an interaction array $\{T_{ij}\}$, define

$$M_i = \mathrm{Max}_j\{T_{ij}\} \qquad (5.22)$$

as the maximum flow out of i.

$$D_i = \sum_j T_{ji} \qquad (5.23)$$

is the total flow into zone i. This is then used to rank order the zones, the highest D_i being the first, and so on. A plot of network topology can then be obtained as follows: if the maximum flow out of i (M_i) is to a lower order centre, which we

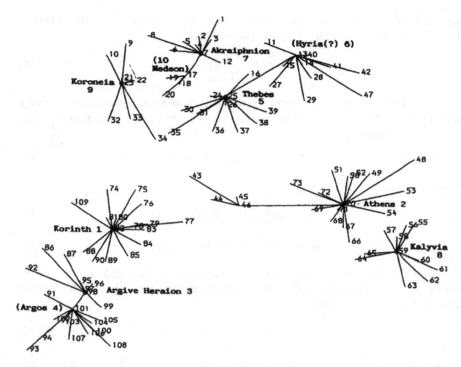

Fig. 5.3 The network structure showing terminals from a model-based analysis of settlement structure in eighth century BC Greece

label J_i, then i is a terminal. Then a vector is plotted connecting each i to J_i. A typical plot is shown in Fig. 5.3.

5.3.5 Rule-Based Representations

There is another style of modelling, currently very fashionable, which makes new kinds of mathematical demands—in this case largely algorithmic. These are agent-based models (ABMs): individual 'agents' act probabilistically (and interact with each other) in and on an environment. There is then the possibility of 'emergent structures'—and we will follow through on this in Chaps. 7 and 8. These are related to, but distinct from, cellular automata models. In this latter case, we would usually have a grid of 'cells', each of which is in a particular 'state', with probabilities of transition to a different state, usually as a function of the states of neighbouring cells. These are sometimes confused with agent-based models but shouldn't be! What they have in common is that they are rules-based.

These kinds of models then link to game theory and computer games which can also be seen as rules-based. Games have 'players' with 'strategies' and can be thought of as a form of agent-based model. From a mathematical point of view, theorems on equilibrium outcomes are important—notably Nash equilibria which, as with other aspects of dynamic modelling, relates to fixed point theorems. We will introduce some urban games in Chap. 7.

5.3.6 Combinations

Most models, of course, combine different elements of the mathematical tool kit. The retail dynamics example of Chap. 3 combines Boltzmann and Lotka-Volterra. Reaction–diffusion models add another dimension, representing phenomena where an entity diffuses across space thereby changes its concentration of density at each point. This turns on a second spatial derivative which is easiest to write down in continuous space form but has to be transformed into discrete space by finite difference methods.

Microsimulation was invented as a means of handling very large arrays which contain a large number of zero entries, building hypothetical populations that 'look like' real ones—based on real probabilities. This is a powerful way of combining data and hence handling missing data issues—but also bringing in a probability approach to modelling.

These approaches can often be seen as alternative 'big picture' perspectives on a particular system of interest—and then may be equivalent at a deeper level. We will show in Chap. 7 that there are versions of the retail model—BLV or ABM— that are more or less equivalent in this respect.

References

Leslie PH (1945) On the use of matrices in certain population mathematics. Biometrika 23:183–212
Nysten JD, Dacey MF (1961) A graph theory interpretation of nodal regions. Papers, Regional Sci Assoc 6:29–42
Rogers A (1973) The mathematics of multiregional demographic growth. Environ Planning 5:3–29
Rogers A (1975) Introduction to multiregional mathematical demography. Wiley, New York
Wilson AG (1974) Urban and regional models in geography and planning. Wiley, Chichester

Further Reading

Artle R (1959) Studies in the structure of the Stockholm economy, the business research unit of the Stockholm school of economics. Cornell University Press, Ithaca, republished 2005
Kim TJ, Boyce DE, Hewings GJD (1983) Combined input–output and commodity flow models for inter-regional development planning. Geog Anal 15:330–342
Rees PH, Wilson AG (1977) Spatial population analysis. Edward Arnold, London

Rogers A (2008) Demographic modelling of migration and population: a multiregional perspective. Geog Anal 40:276–296
Stone R (1967) Mathematics in the social sciences. Chapman and Hall, London
Stone R (1970) Mathematical models of the economy. Chapman and Hall, London
Wilson AG (2006) Ecological and urban systems models: some explorations of similarities in the context of complexity theory. Environ Planning A 38:633–646
Wilson AG (1970) Entropy in urban and regional modelling. Pion, London, Chapter 3

Chapter 6
Adding Depth-1: Spatial Interaction and the Location of Activities

6.1 Introduction

We have already seen, for example through the retail model in Chap. 2, that interaction and location are critical components of urban and regional analysis. In this chapter, we explore the basic design decisions that have to be made in different sectors of application—whether transport, retail, public services or residential location. When we complete this toolkit by adding dynamics in Chap. 7, we will be in a position to rewrite the classical models of agriculture, industry, residential and central places. We have already seen that there is a wide range of application in other disciplines.

We proceed by reviewing in turn the main model design decisions (6.2), the range of application (6.3) and the derivation of the core interaction model (6.4). We conclude by briefly demonstrating the depth and effective of the application of the core model through the experience of GMAP Ltd (6.5).

6.2 Model Building Design Decisions

For any particular system of interest, as we have seen, we would usually represent spatial interaction through a discrete zone system (as shown, for example, in Fig. 1.2 in Chap. 1).

It is always necessary to bear in mind that a means must be found of 'closure'. For any given system of interest, there will be flows across the external boundary and these should be captured by the addition of 'external zones' (which can become larger at greater distances from the system of interest). We then have to decide on the level of detail in characterising flows—technically, the level of disaggregation. We will have to decide whether we can separately model the total number of trips produced in each zone—the origins—and the total number of trips attracted to each zone—the destinations. In the retail example, only the trip origin totals in the form of total expenditure were assumed known and in that case, the totals attracted into each destination could be calculated, this making the model, as we have seen, also function as a location model through predicting these totals.

In some cases, we will seek to integrate the interaction model with other frameworks—inter-regional flows in an input–output model for example.

6.3 Applications

6.3.1 The Range

A first objective is to achieve the understanding which represents an element of our knowledge of the basic science of the city. When we are confident in our models—perhaps tested through 'back-casting'—they can then be deployed for forecast-ing—predicting the consequences of changes in exogenous variables. In particular, they can be used in public or private planning contexts on a 'What if ?' basis. If a store is opened in a particular place, what revenue does it attract (and from what other stores is this drawn?)? More generally, different plans or scenarios can be tested. In effect, the models become the basis of the planning equivalent of a flight simulator. This kind of forecasting can be very effective for the short run—for example in planning retail networks—but may be more problematic for the longer run because of the effects of nonlinearities. We take this further in Chap. 7.

6.3.2 Interpretations

If the models are used to forecast flows, some measures are needed—indicators—of the effectiveness of a plan. This can be done formally through cost-benefit analysis in many circumstances using, for example, measures of consumers' sur-plus as part of the 'benefit'. This will be pursued in more detail below. In a private sector example such as retail, the key measure and aim will be profit maximisation.

We will see that the spatial interaction models can be related more formally to economics. When they are represented in mathematical programming form, the dual programme can be constructed and this provides measures of comparative advantage through rents and utilities.

6.4 Derivations of the Core Spatial Interaction Model

6.4.1 Entropy Maximising

An important means of generating spatial interaction models is that of entropy maximisation. This shifts the underlying idea from 'gravity' to a form of statistical averaging. The core 'entropy' idea is to count the number of possible 'system states' that are consistent with 'constraints' that represent our known knowledge of the system of interest. We begin by developing the so-called doubly-constrained model within this formalism which will illustrate the basic idea.

We use a general notation that is common in the transport field. We have a zone system with labels i or j; T_{ij} is the flow of trips (let us say) from i to j; O_i is the total number of trips from i and D_j the total number terminating in j. c_{ij} is a measure of impedance between i and j—most simply distance, but a term which, as we have seen, can be developed as a 'generalised cost'.

We can begin by noting that a gravity model might then be

$$T_{ij} = KO_iD_jc_{ij}^{-\beta} \tag{6.1}$$

with, if we pressed the analogy, $\beta = 2$, though there is no reason why this should be the case in a social science system of interest. However, the $\{T_{ij}\}$ calculated from such a model cannot satisfy our prior knowledge:

$$\Sigma_j T_{ij} = O_i \tag{6.2}$$

$$\Sigma_i T_{ij} = D_j \tag{6.3}$$

Historically, factors were added pragmatically to give:

$$T_{ij} = A_iB_jO_iD_jc_{ij}^{-\beta} \tag{6.4}$$

with

$$A_i = 1/\Sigma_k B_k D_k c_{ik}^{-\beta} \tag{6.5}$$

and

$$B_j = 1/\Sigma_k A_k O_k c_{kj}^{-\beta} \tag{6.6}$$

We now ask a new question: how many possible states of the system are compatible with the constraints? The answer is

$$W = T!/\Pi_{ij}T_{ij}! \tag{6.7}$$

where T is the total population. Boltzmann's great discovery for the equivalent problem in physics was that $\{T_{ij}\}$ which resulted from maximising W in (6.7) subject to constraints (6.3) and (6.4) (and one other constraint to be introduced shortly) was overwhelmingly the most probable. The additional constraint, equivalent to the physicist's energy constraint, is

$$\Sigma_{ij}T_{ij}c_{ij} = C \tag{6.8}$$

representing, formally, a known total cost of travel. In practice, this does not have to be known as such: the associated parameter, β, can be estimated through model calibration.

In practice, we maximise logW subject to (6.2), (6.3) and (6.8). This gives

$$T_{ij} = A_iB_jO_iD_j\exp(-\beta c_{ij}) \tag{6.9}$$

$$A_i = 1/\Sigma_k B_k D_k \exp(-\beta c_{ik}) \tag{6.10}$$

$$B_j = 1/\Sigma_k A_k O_k \exp(-\beta c_{kj}) \tag{6.11}$$

Note that the power function in the gravity model has now become an exponential function. However, if we replace c_{ij} in (6.8) by $\log c_{ij}$ and redo the maximisation, we get a power function again from the entropy maximising approach. This suggests that if an exponential function fits best, cost is perceived as linear, if a power function fits best as logarithmic—representing shorter and longer average trips respectively. This interpretation represents an intuitively plausible idea in relation to the perception of trips. More generally, if there is a function that fits the data better, a suitable form of equation (6.8) can be used to generate that function for the model and this can then be interpreted in relation to the perception of travel cost.

These models are know as entropy maximising models because

$$S = k\log W \tag{6.12}$$

is equivalent to the entropy that turns up in physics (and is engraved on Boltzmann's gravestone in Vienna). It is also equivalent to Shannon's measure of 'information' and relates to Bayes' theorem in statistics.[1] (The story is told that Shannon consulted the famous John von Neumann on what he should call his measure and von Neumann replied that he should call it 'entropy' because of its relationship to the quantity that appears in physics, but also because no one would understand it which meant that in any argument, he would always win!) Jaynes (1957) demonstrated its wide applicability.

There is a crucial lesson in this formulation, already hinted at: our knowledge of the system of interest is represented in the constraints. The model, therefore, represents the 'most likely' system state which is consistent with the constraints. It is also optimally 'blurred' relative to a related linear programme. Recall that as $\beta \to \infty$ in (6.9), then $\{T_{ij}\} \to$ the LP solution—a feature we will explore further below. This 'lesson' means that we can enrich the core model by developing submodels for O_i and D_j and exploring a range of functions for the impedance, c_{ij} together with the idea of 'generalised cost'. We will see later, when we disaggregate, that, we can add transport modes and indeed other features and we can load flows on to real networks and hence model congestion. But first, we note other variants of the core model.

---

[1] See Wilson (1970), p. 9.

6.4.2   A Family of Models

The basic 'family of models' was introduced in Sect. 5.2.4 of Chap. 5 and is created by dropping one or more of the constraints (6.2) and (6.3). If we drop (6.3) for example, the model becomes

$$T_{ij} = A_i W_j \exp(-\beta c_{ij}) \tag{6.13}$$

With

$$A_i = \Sigma_k W_k \exp(-\beta c_{ij}) \tag{6.14}$$

In this case, an attractiveness factor W_j has been added and we can now calculate model-predicted inflows:

$$D_j = \Sigma_i T_{ij} \tag{6.15}$$

so that the model functions as a location model. This, for obvious reasons is known as the origin (or production) constrained model and it is, of course, the retail model in another notation. The mirror image is the destination (or attraction) constrained model which is used, for example, for residential location modelling in relation to workplaces. If both constraints are dropped, then something like the original gravity model is recovered. If both are applied, we have the doubly constrained model used to illustrate the derivation above. There are situations where it is necessary to develop a hybrid model—part doubly constrained, part singly constrained.

Consider the case of student flows: the O_is are students by residence and the D_js are university places. There will be a set, Z_1 say, of universities that are 'full'—i.e. constrained, and one, Z_2, that are unconstrained. We would then have something like

$$T_{ij} = A_i B_j O_i D_j \exp(-\beta c_{ij}), \ j \, \varepsilon \, Z_1 \tag{6.16}$$

$$T_{ij} = A_i O_i W_j \exp(-\beta c_{ij}), \ j \, \varepsilon \, Z_2 \tag{6.17}$$

the problem is that the balance between Z_1 and Z_2 can change during the doubly-constrained iteration procedure and so there has to be an outer iteration. However, the moral is: a well-defined problem can be handled!

Another illustration of this principle comes in modelling trade flows. In that case, the zonal input–output equations can be added as constraints and this produces and interesting hybrid model.

6.4.3   Indicators

We consider in turn: accessibility, consumers' surplus, catchment populations, and effectiveness and efficiency indicators. To explore accessibility, we return to the retail model and consider

$$Q_i = \Sigma_j W_j \exp(-\beta c_{ij}) \tag{6.18}$$

Intuitively, it can be seen that this is a measure of accessibility to shops for residents of zone i. It can be used in a similar way in other contexts. As a concept, it has been around for a long time possibly originating in Hansen (1959).

To explore consumers' surplus, we again use the retail model whose main equation is repeated here for convenience.

$$S_{ij} = A_i e_i P_i W_j^\alpha \exp(-\beta c_{ij}) \tag{6.19}$$

We can write this as

$$S_{ij} = A_i e_i P_i \exp\{\beta[(\alpha/\beta)\log W_j - c_{ij}]\} \tag{6.20}$$

so

$$[(\alpha/\beta)\log W_j - c_{ij}] \tag{6.21}$$

can be taken as a measure of benefit, and

$$\Sigma S_{ij}[(\alpha/\beta)\log W_j - c_{ij}] \tag{6.22}$$

as a measure of consumers' surplus.

To explore catchment populations, we introduce a new element of notation: let an asterisk denote summation, so

$$S_{i*} = \Sigma_j S_{ij} \tag{6.23}$$

and then

$$\Pi_j = \Sigma_i (S_{ij}/S_{i*})P_i \tag{6.24}$$

is a measure of the catchment population of j. Note that:

$$\Sigma_j \Pi_j = \Sigma_i P_i \tag{6.25}$$

which is a useful degree of consistency.

We can now use (6.24) as a denominator to calculate performance indicators. For a retail centre, j, D_j/Π_j can be taken as the revenue per head of catchment population—a useful efficiency indicator. But now we can introduce a new one. Define

$$\Omega_i = \Sigma_j (S_{ij}/S_{*j})W_j \tag{6.26}$$

which is the proportion of W_j 'used' by residents of i and so Ω_i/P_i is a measure of effectiveness of service delivery. Note then that a service can be efficient but not necessarily effective in delivery—consider, e.g. National Health Service dental practices in the UK.

6.4.4 Disaggregation

So far, the argument has been presented in terms of an aggregate model. Nearly always, it is necessary to disaggregate to achieve an appropriate degree of realism. This is usually straightforward but can generate some new insights as well. Recall the suggested retail disaggregation in Chap. 1 which included the introduction of a generalised cost with several elements—a form of disaggregation and a set of Ws representing different elements of attractiveness. But now let us illustrate disaggregation further with the transport model. Suppose we add mode k and person type n (e.g. car owner/non car owner) as superscripts. We then have

$$T_{ij}^{kn} = A_i^n B_j^n D_j^n \exp(-\beta c_{ij}^k) \tag{6.27}$$

However, we may want to separate distribution and modal split (again using the notation that an asterisk replacing a subscript or superscript denotes summation over that index):

$$T_{ij}^{*n} = A_i^n B_j^n O_i^n D_j^n \exp(-\beta C_{ij}^n) \tag{6.28}$$

and then modal split M_{ij}^{kn} defined by

$$T_{ij}^{kn} = T_{ij}^{*n} M_{ij}^{kn} = T_{ij}^{*n} \exp(-\lambda^n c_{ij}^k)/\Sigma_k \exp(-\lambda^n c_{ij} k) \tag{6.29}$$

We now have two key parameters, β and λ, and to make (6.28) and (6.29) consistent with (6.27), we have

$$\exp(-\beta C_{ij}^n) = \Sigma_k \exp(-\beta c_{ij}^k) \tag{6.30}$$

which shows how we can construct a generalised cost for distribution from modal costs.

6.4.5 Mathematical Programming Formulations

We have already introduced the transportation problem of linear programming (LP) in Chap. 5. Here we note the remarkable result—proved by Evans (1973)—that this is the $\beta \to \infty$ limit of the doubly constrained model. Recall the form of the LP model:

$$\text{Min } C = \sum_{ij} T_{ij} c_{ij} \tag{6.31}$$

subject to

$$\sum_j T_{ij} = O_i \tag{6.32}$$

and

$$\sum_i T_{ij} = D_j \qquad (6.33)$$

We can then note that the EM model is constructed by maximising an entropy function, $-\sum_{ij}T_{ij}\log T_{ij}$, with the LP objective function now turned into a constraint:

$$\sum_{ij} T_{ij}c_{ij} = C \qquad (6.34)$$

where, of course, C will have a value greater than the LP C in order to produce a finite β.

This preliminary analysis then enables us to explore duals and we do this first in relation to the transportation problem in (6.31)–(6.33) above. The dual in this case is formed by introducing variables associated with the constraints and formulating an appropriate objective function. In this case, let μ_i be the variable associated with the O_i constraint and v_j that with the D_j one. The dual problem is then:

$$\text{Max } C' = \sum_i \mu_i O_i + \sum_j v_j D_j \qquad (6.35)$$

such that

$$c_{ij} - \mu_i - v_j \geq 0 \qquad (6.36)$$

μ_i and v_j can be interpreted as measures of comparative advantage—or indeed as 'rent'—because this is what can be extracted by a landlord from the 'advantage'.

We can now ask the question: what is the dual of the nonlinear EM programme which produced the doubly constrained model. We can introduce dual variables μ_i and v_j as before but now an additional dual variable, β, associated with the cost constraint. The dual turns out to be

$$T_{ij} = \exp(-\mu_i - v_j - \beta c_{ij}) \qquad (6.37)$$

which can be easily converted to the standard form—showing that the A_i and B_j factors are related to the dual variables and hence 'rents'.

6.4.6 *Random Utility Theory and Related Economic Models*

Economists, on the whole, have not liked entropy-maximising models! We have indicated that it is easy to convert an 'optimising' economic model—the Herbert-Stevens model being a good example (which we explore in more detail in Chap. 9)—into a more realistic 'imperfect market model' by converting to EM; and this has the added advantage that terms such as A_i and B_j can be given an economic interpretation. The economists' response has been to employ a different form of 'dispersion'—deviation from the optimum. This has been done in micro economics by assuming that some dispersion has to be added to the utility functions that would normally be maximised on a 'rational person' basis. This produces different variants of random utility theory and generates the much used logit model. The usual method is to assume an average utility function and then to add a random component. By making a suitable assumption about the distribution of the random component, a model can be derived—in probability terms, for example:

$$p_{ij} = \exp(\beta U_{ij})/\Sigma_{ij}\exp(\beta U_{ij}) \tag{6.38}$$

with suitable definitions of U_{ij}, this can be transformed into an EM model. The assumed 'random distribution', of course, has to be closely related to the entropy function to get a plausible answer!

6.4.7 Other Formulations

There are many alternative derivations of the core model but these will not be pursued in detail here. They include other approaches from statistical mechanics such as the Darwin-Fowler method; Bayesian statistics and maximum likelihood (which is useful for model calibration); Shannon's information theory, based on the entropy of a probability distribution. Batty (1974) introduced the concept of spatial entropy; Snickars and Weibull (1977), the addition of prior information. A different approach, the intervening opportunities was introduced by Stouffer (1940) and developed by Schneider (1967) (which can be converted into an entropy maximising form by suitable definitions of the constraints); and Fotheringham (1983) produced the idea of competing destinations. Smith and Hsieh (1997) used a Markov formulation; and Haag (1990) a master equations' one.

Since many of these alternatives are more or less equivalent, it is appropriate for a practioner to work with a formulation that is familiar so that it is always possible to be able to derive a good core model.

6.5 Testing

The kind of spatial interaction models described in this chapter have been extensively tested, particularly in the transport and retail sectors. Highly disaggregated models have been developed and used to estimate the costs and benefits of alternative transport investments or the optimisation of networks of retail outlets. They can also be used in market analysis, as the basis of decision support systems and for customer relationship management (CRM). They have not been used as extensively to help optimise the location of public facilities.

A good example of the deployment of these systems was in the development of GMAP Ltd as a University of Leeds spin-out by Clarke and Wilson in the late 1980s and 1990s. It provides an interesting case study in the development of such businesses and the exploitation of university research and intellectual property—though in this case, there were no patents, only high levels of skill that initially it was difficult for others to replicate.

GMAP's skills were rooted in the core interaction model but also encompassed small area demand forecasting (for example, in retail, finance and medical) and measures of the attractiveness of supply outlets. This allowed revenue (or usage) predictions to be made and provided a 'What if' forecasting capability for clients. The client base was 'retail' in the broadest sense, ranging from shops and shopping centres to finance, auto dealerships and garages. There was a small amount of public sector work, for example in planning (assessing the impact of out-of-town centres), health, water and education. There was also a small amount of work for industry clients—in chemicals and pharmaceuticals for example. Clients included Smith, Toyota, Tesco, Ford, Storehouse, Mansfield Breweries, Whitbreads, BP and Mobil, Smith Klyne Beecham, and the Halifax Building Society.

The company was built on a DIY basis with no capital investment. In the first three or four years, turnover grew from a tiny £20 k to perhaps 100 k but then grew steadily, exceeding £1 M p.a. in 1990—when GMAP was formed as a company—until at its peak it was around £6 M p.a. with over 100 employees. In 1997, the automotive part of GMAP was sold to Polk, an American company, and the residual to the Skipton Building Society, as part of their group of marketing companies, in 2001.

6.6 Concluding Comment

Spatial interaction modelling is an effective tool for modelling flows and levels of activity attracted to locations. It has been deployed in a variety of forms and thoroughly tested. It is usually the case that an appropriate model has to be developed on a bespoke basis for any particular system of interest or planning or business problem but the principles of model design described in this chapter can be brought to bear in each case.

References

Batty M (1974) Spatial entropy. Geog Anal 6:1–31
Evans SP (1973) A relationship between the gravity model for trip distribution and the transportation model of linear programming. Transp Res 7:39–61
Fotheringham AS (1983) A new set of spatial interaction models: the theory of competing destinations. Env Planning A 15:15–36
Haag G (1990) Master equations. In: Bertuglia CS, Leonardi G, Wilson AG (eds) Urban dynamics: designing an integrated model. Routledge, London, pp 69–83 new edition, 2011

Hansen WG (1959) How accessibility shapes land use. J Am Inst Plan 25:73–76
Jaynes ET (1957) Information theory nand statistical mechanics. Phys Rev 106:620–630
Schneider M (1967) Direct estimation of traffic volume at a point. Highway Res Record 165:108–116
Smith TE, Hsieh SH (1997) Gravity-type interactive Markov models—part I: a programming formulation for steady states. J Regional Sci 37:683–708
Snickars F, Weibull JW (1977) A minimum information principle. Regional Sci Urban Econ 7:137–168
Stouffer SA (1940) Intervening opportunities: a theory relating mobility and distance. Am Sociological Rev 5:845–867

Further Reading

Birkin M, Clarke GP, Clarke M, Wilson AG (1996) Intelligent GIS: location decisions and strategic planning. Geoinformation International, Cambridge
Birkin M, Clarke GP, Clarke M (2002) Retail geography and intelligent network planning. Wiley, Chichester
Clarke GP, Wilson AG (1987) Performance indicators and model-based planning II: model-based approaches. Sistemi Urbani 9:138–165
Wilson AG (1970) Entropy in urban and regional modelling. Pion, London
Wilson AG (2000) Complex spatial systems. Prentice Hall, New Jersey, Chapter 6
Wilson AG (2010) Entropy in urban and regional modelling: retrospect and prospect. Geog Anal 42:364–394

Chapter 7
Adding Depth-2: Structural Dynamics

7.1 The Retail Model as an Example

We now return to the retail model introduced in Chap. 2 which is repeated here for convenience. Recall that e_i is the per capita expenditure by each of the P_i residents of zone i and W_j is the size of a centre, and by raising it to a power, α, it becomes a measure of the attractiveness of retail centre j. If $\alpha > 1$, this will represent positive returns to scale for retail centres. c_{ij} is a measure of travel cost as usual.

$$S_{ij} = A_i e_i P_i W_j^\alpha \exp(-\beta c_{ij}) \qquad (7.1)$$

where

$$A_i = \Sigma_k W_k^\alpha \exp(-\beta c_{ik}) \qquad (7.2)$$

to ensure that

$$\sum_j S_{ij} = e_i P_i \qquad (7.3)$$

We can use the model to calculate the total revenue attracted to a particular j:

$$D_j = \Sigma_i S_{ij} \qquad (7.4)$$

which is, substituting from (7.1) and (7.2)

$$D_j = \Sigma_i [e_i P_i (W_j)^\alpha \exp(-\beta c_{ij}) / \Sigma_k (W_k)^\alpha \exp(-\beta c_{ik})] \qquad (7.5)$$

We showed that a simple hypothesis for the dynamics is:

$$\Delta W_j(t, \ t+1) = \varepsilon [D_j(t) \ - \ KW_J] W_j(t) \qquad (7.6)$$

We saw that this is a form of Lotka–Volterra equation. The expression $[D_j(t) - KW_j(t)]$ is a measure of 'profit' (or 'loss'). So Eq. 7.6 is representing a

A. Wilson, *The Science of Cities and Regions*, SpringerBriefs in Geography, DOI: 10.1007/978-94-007-2266-8_7, © The Author(s) 2012

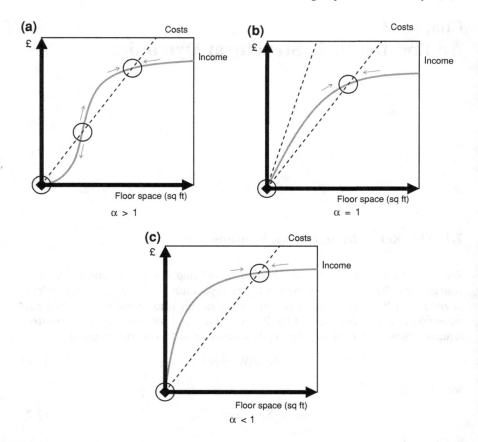

Fig. 7.1 Revenue-costs graphs for a zone

hypothesis that if a centre is profitable, it will grow; otherwise, it will decline. The parameter ε measures the strength of response to this signal.

At equilibrium, $\Delta W_j(t, t + 1)$ is zero, so

$$D_j = C_j = KW_j \qquad (7.7)$$

That is

$$D_j = \Sigma_i \{e_i P_i W_j^\alpha \exp(-\beta c_{ij}) / \Sigma_k W_k^\alpha \exp(-\beta c_{ik})\} = k_j W_j \qquad (7.8)$$

We now have to seek some deeper understanding of what this mechanism means. We plot D_j as a function of W_j and the straight line, KW_j: the income-costs zone graph in Fig. 7.1. We show three cases as (a), (b) and (c). These correspond to $\alpha > 1$, $\alpha = 1$ and $\alpha < 1$.

In case (a) there are three points of intersection—the equilibrium points—two stable (marked by circles) and one unstable (the centre of the three circles); in case (b) the shape of the D_j curve is different because it has a finite gradient at the origin and there are two points of intersection, both stable; in case (c) the gradient of the D_j curve is infinite at the origin and there are two points of intersection with only the non-zero one stable. In cases (a) and (b), the gradient of the KW_j line may be sufficiently steep that there is no non-zero intersection. We can therefore summarise the possible outcomes for this zone as follows:

$\alpha > 1$:

a. If there are two intersections, then W_j can be either 0 or finite; this has been described as the d-p (development possible) state. Which state is adopted depends on the initial conditions: if the initial value is below the unstable equilibrium point, the zone will move to zero—the n-d-p state, and vice versa.
b. If there is no non-zero intersection, then W_j can only be zero—the n-d-p state.

$\alpha = 1$: as for $\alpha > 1$, but if there is a finite intersection, then the d-p state will be the outcome unless $W_j = 0$ initially.

$\alpha < 1$: in this case, the stable state is always non-zero so the zone will always be in the d-p state.

This analysis offers some insight into how this system displays the generic characteristics of a nonlinear system. The existence of a zero stable state in the $\alpha \geq 1$ cases illustrates *multiple equilibria*—a very large number of possibilities when applied to a large multi-zoned system. It illustrates *path dependence*,[1] because of the dependence of the particular solution on the initial conditions—and path dependence, as we have seen, arises from a sequence of initial conditions. And it illustrates *phase changes*, because if, for example, the gradient of the KW_j line increases and the finite equilibrium point disappears, the zone could jump from a non-zero state to a zero one.

The future possible development of a system may well be strongly dependent on initial conditions and path dependence. The initial conditions at a point in time can be thought of as the 'DNA' of the system. For the retail model, these initial conditions are

$$[\{e_i\}, \{P_i\}, \{W_j\}, \{c_{ij}\}, \alpha, \beta] \tag{7.9}$$

These arrays are all exogenous except for $\{W_j\}$. However, as we have noted, the particular equilibrium which becomes the governing one at a point in time is determined by the positions of the initial W_j for each zone relative to the unstable equilibrium. (This would be important in relation to any investment in a new centre: there is a need to ensure that the investment takes W_j above that unstable point.) See Fig. 7.2.

If we add some stochastic features representing possible future values of exogenous variables, this allows us to develop a 'cone of possible development' in

---

[1] An idea introduced by Arthur (1988).

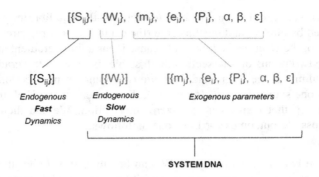

$$[\{S_{ij}\}, \{W_j\}, \{m_j\}, \{e_i\}, \{P_i\}, \alpha, \beta, \varepsilon]$$

$[\{S_{ij}\}]$ $[\{W_j\}]$ $[\{m_j\}, \{e_i\}, \{P_i\}, \alpha, \beta, \varepsilon]$

Endogenous   *Endogenous*   *Exogenous parameters*
Fast   **Slow**
Dynamics   *Dynamics*

SYSTEM DNA

Fig. 7.2 Path dependence and 'DNA'

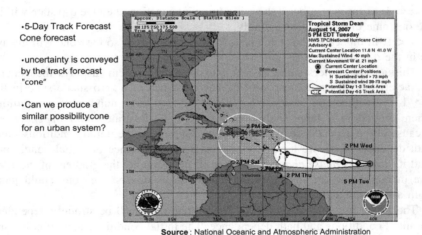

•5-Day Track Forecast
Cone forecast

•uncertainty is conveyed
by the track forecast
"cone"

•Can we produce a
similar possibility cone
for an urban system?

Source : National Oceanic and Atmospheric Administration

Fig. 7.3 The 'cone of possible development' for a hurricane

a high dimensional configuration space. This is illustrated in two dimensions in Fig. 7.3—the method used by weather forecasters to chart the possible path of a hurricane.

In Fig. 7.4, we show how in principle to build a possibility cone, this time shifting to a notional three dimensions.

The South Yorkshire study area and the 'root DNA' is shown in Fig. 7.5.

A base South Yorkshire model run of the conventional model is shown in Fig. 7.6.

This example was motivated by the fact that Rotherham town centre was in decline (as well as that there were far fewer retail centres than in London!). This encouraged us to extend the DNA idea via 'genetic medicine' to 'genetic

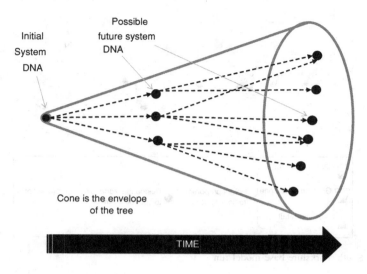

Fig. 7.4 Building a possibility cone

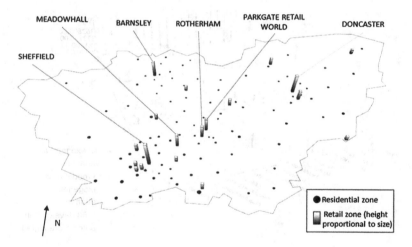

Fig. 7.5 The South Yorkshire study area

planning': what would have to happen to enable the High Street to compete with out-of-town shopping centres? What changes do we need to make to the DNA that ensures this centre is more stable? We can illustrate a response with two ideas: first, the use of parallel coordinates in visualisation; and secondly, the tweaking of the DNA, in this case by reducing the costs to Rotherham town centre, say by

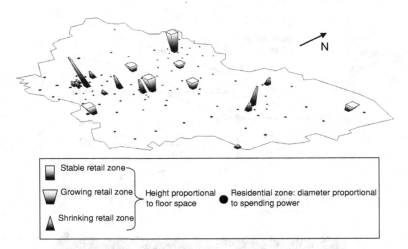

Fig. 7.6 South Yorkshire base model run

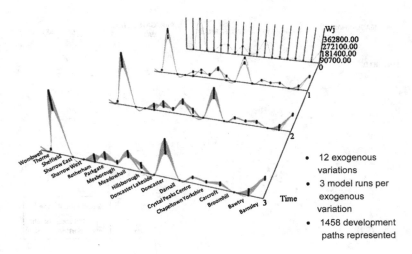

Fig. 7.7 A parallel coordinates representation

reduced parking charges. The parallel coordinates representation is shown in Fig. 7.7—in effect a probability cone in 19 dimensions.

In Figs. 7.8 and 7.9, we show the effect on Rotherham centre of reducing the travel costs. This shows how it comes alive again relative to its Parkgate out-of-town competitor.

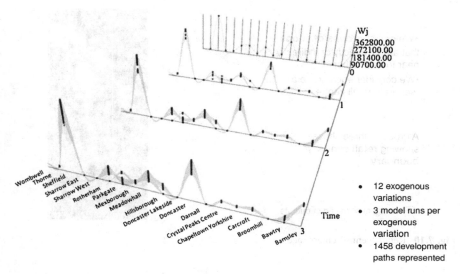

- 12 exogenous variations
- 3 model runs per exogenous variation
- 1458 development paths represented

Fig. 7.8 Reducing travel costs to a city centre

Fig. 7.9 Before and after

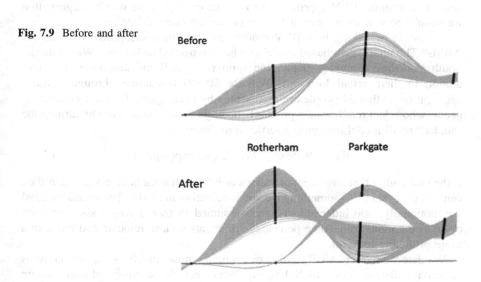

7.2 Agent Based Modelling Approaches

A common approach in contemporary complexity science is agent-based modelling. Each agent is some kind of individual who is assigned rules of behaviour which represent the essence of the dynamics. In suitable circumstances, we will

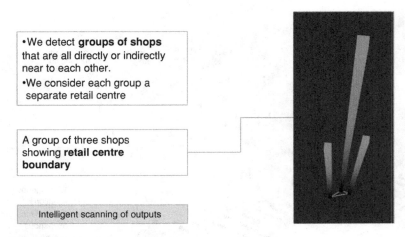

•We detect **groups of shops** that are all directly or indirectly near to each other.
•We consider each group a separate retail centre

A group of three shops showing **retail centre boundary**

Intelligent scanning of outputs

Fig. 7.10 Retail 'centre' emergence

find emergent structures. An interesting question is whether such an approach is consistent with the BLV approach. We offer an example here which suggests that we should be able to achieve at least approximate equivalence.

We begin by asking how BLV models can inform development of rules for ABMs? The retail agent based model can be constructed as follows. We retain the South Yorkshire example—population approx. 1.2 million—and assign the population to their actual locations. We use 50,000 consumers through a mini-aggregation so that 24 people are represented by each agent. They are consuming agents who select retailers on a probabilistic basis which we develop by turning the standard retail model into probabilistic form. Suppose

$$R_{ij} = W_j^\alpha \exp(-\beta c_{ij})/\Sigma_k W_k^\alpha \exp(-\beta c_{ik}) \tag{7.10}$$

is the probability that a resident of i shops in j. Note that we have not yet identified centres, only 'shops'—something we are anticipating in (7.10). The retails 'agents' are, notionally, 500 independents, each assumed to own a single shop. In each cycle of the model run, some percentage of agents seek to relocate and move to a more profitable location.

We detect emergent retail *centres* by grouping those 'nearby'—say on the basis of walking distance—within 200 m, say. We detect closed groups of shops where each member is near to at least one other member of that group. A group of three shops showing a retail centre boundary is shown in Fig. 7.10.

In this way, we can identify separate emergent retail centres. The outcome for the South Yorkshire example is shown in Fig. 7.11 and this is plausibly near the results from the equivalent BLV model (Fig. 7.6).

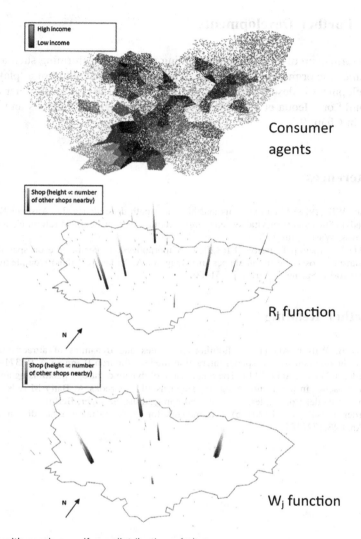

Fig. 7.11 An ABM outcome

This illustrates an important feature of model building: it is worth exploring apparently very different approaches to explore possible equivalences at a deeper level.

7.3 Further Developments

The formalism of agent-based modelling suggests transforming such a model into a game: one or more retail agents can be put under the control of a 'player'. Such a development is described in Dearden and Wilson (2011a, b). Another approach is to build on Medda et al. (2009) work on morphogenetic models and we explore this in Chap. 9.

References

Arthur WB (1988) Urban systems and historical path dependence. In: Ausubel JH, Herman R (eds) Cities and their vital systems: infrastructure past, present and future. National Academy Press, Washington DC
Medda F, Nijkamp P, Reitveld P (2009) A morphogenetic perspective on spatial complexity: transport costs and urban shapes. In: Reggiani A, Nijkamp P (eds) Complexity and spatial networks. Springer, Berlin, pp 51–60

Further Reading

Harris B, Wilson AG (1978) Equilibrium values and dynamics of attractiveness terms in production-constrained spatial-interaction models. Environ Planning A 10:371–388
Dearden J, Wilson AG (2011a) The relationship of dynamic entropy maximising and agent based approaches in urban modelling. In: Heppenstall A, Crooks A, Batty M (eds) Spatial agent based models: principles, concepts and applications. Springer, Berlin
Dearden J, Wilson AG (2011b) A framework for exploring urban retail discontinuities. Geog Anal 43:172–187

Chapter 8
Tools for Complexity Science

8.1 The 'Complexity' Thesis

In this chapter, we pose the question: what does the science of cities, and in particular, the mathematical modelling of cities, offer complexity science? A prior question, of course, is: what is complexity science? Complex systems are characterised by needing many variables to describe them and having strong interdependencies between the elements of the system. When represented mathematically, these interdependencies will typically be nonlinear relationships. Obvious examples of complex systems are human beings, brains, economies, ecosystems, languages and, of course, cities and regions. From the perspective of this book, the 'systems, theory and methods' framework is an important foundation. Everything we have asserted about cities and regions in this respect repeats itself for any complex system: the system of interest must be well defined, we must have a theory of how it works, and we must have a tool kit of methods to facilitate building models that represent the theory. This raises all the issues of scale and the interactions between different scales.

It can be argued, perhaps with some exaggeration, that the most important and currently interesting science is about complex systems. Huge claims are made for this interdisciplinary perspective and, as a field, it is the height of research fashion. It represents a shift from the highly successful reductionist mode of recent decades in fields such as molecular biology and elementary particle physics, to the study of the properties of whole systems. It is useful to recall Weaver's extraordinary prescient classification of problems and systems: from the simple (Newtonian, for example, small numbers of variables), through systems of disorganised complexity (Boltzmann, for example, large numbers of variables but weak interactions), to systems of organised complexity (large numbers of variables, strong interdependencies). The most challenging problems of complexity science are in the third category though much of the tool kit relates to the first two. We can argue in relation to cities and regions that these are certainly complex enough to be interesting, but not as complex as biological systems, for example, and so it may be possible to make more rapid progress. It is helpful, therefore, to review our

A. Wilson, *The Science of Cities and Regions*, SpringerBriefs in Geography,
DOI: 10.1007/978-94-007-2266-8_8, © The Author(s) 2012

capabilities for developing theory-building and modelling in the urban and regional field and see what we might offer the science of complex systems more broadly; and vice versa: are there elements of the complexity science tool kit that we have not yet exploited? Key concepts of complexity science include systems, hierarchies and scales, emergence and the evolution of order and nonlinear mathematics. In the next section, we review our capabilities and respond to the questions posed.

8.2 New Challenges, New Tools

Most of the mathematical tools that are necessary for complexity science are probably already available to us. There is a challenge embedded in this statement: that the mathematicians who work at the research front line and know these fundamentals in depth do not typically work on real and messy systems. It is necessary, therefore to bridge a gap between this depth of knowledge in mathematics and its application. The elements of the tool kit, as we saw in Chap. 5, can perhaps be characterised as:

- algebra, matrix algebra and tensor algebra
- basic calculus
- optimisation and mathematical programming
- nonlinear dynamical systems analysis
- simulation methods when analytical solutions (or system representations) are not available.

The properties of complex systems that have been of most interest, again as we have seen illustrated, are phase changes, path dependence—that is dependence on a sequence of initial conditions—and the emergence of structure. What seems to have happened is that complexity scientists have focused on systems that exhibit these properties, often artificial, or ones that are relatively well defined subsystems—but which neglect the interdependencies with other subsystems that makes real systems challenging and interesting. These practices have created whole subdisciplines: 'artificial life', for example, or 'network science'. What is needed is a fuller engagement of the complexity scientists and mathematicians who have the tools with those who work on the real systems, whether in biology, physics, engineering, business or urban and regional science.

From this preliminary and very broad analysis, can we begin to chart in broad terms where the new research directions will emerge from? Indications have already been given in earlier chapters but here we attempt a summary as a preliminary to charting some specific examples in the next chapter. We have to think through the two-way process: what can complexity scientists learn from urban and regional science? And, vice versa?

In the first category, there is undoubtedly the treatment of space. The representation of spatial structures with different kinds of discrete zone systems seems to provide the basis for better modelling tools than are used in some other 'complexity' fields. It may also be that there are nonlinear urban and regional models within which it has been possible to articulate in some detail the mechanisms of system evolution that may point up research directions for other systems.

In the second category, it is almost certainly the case that urban and regional modellers have restricted themselves to particular kinds of equations that generate structure—particularly those connected to what might be called the Lotka-Volterra family and a wider search of forms of Kolmogorov or Fokker-Planck equations might provide new models of spatial structural evolution. It is also the case that many of the models in complexity science—network science being an excellent example—have been developed without knowledge of similar models from earlier decades. In the network science case, for example, the field, although huge, is impoverished because it is unaware of the possibility of using spatial interaction models to estimate flows that can then be loaded on to networks—generating weighted networks which are a much richer basis for analysis than that provided by the topology of nodes and links.

8.3 Will Complexity Science Succeed?

The idea of complexity science as described in broad terms here is helpful in at least two ways. First, it helps to provide a broader tool kit for modellers of complex nonlinear systems. Secondly, it encourages a shift of focus in science from the reductionist to the system. In these terms it will be important and it will succeed. However, whether complexity science is best organised in specialist centres and institutes rather than within the centres and departments that have particular complex systems as central to their interests is debateable. This turns on the broad question: is there a possibility that complexity science can offer a general theory of systems? This kind of programme has been launched before—for example through von Bertalanffy's work in the 1950s and the development of so-called 'general systems theory'. But this programme was not sustained. Probably a balance has to be struck: complexity science is helpful in terms of the tool kit and the focus—as indicated above. It is probably also helpful that there are people who work on the general characteristics of complex systems. Holland (1995, 1998) is a striking example with his general definition of 'agents' and 'environments'. It can also, in this guise, help to lay a high-level foundation for interdisciplinarity. John Casti (1995) provides a good summary which positions complexity science well: "......the creation of a science of complex systems is really a subtask of the more general, and much more ambitious programme of creating a theory of models. Complexity—as a science—is merely one of the endless rungs on this ladder".

Concepts introduced in earlier chapters—interdependencies, combining input–output, BLV and reaction–diffusion models, linking scales through layered models, cascading catastrophes and deepening the mathematics (stability and instability in high dimensional manifolds, exploring the full range of mathematical 'systems' that will create structure) are all elements of complexity science. The wider literature is now voluminous. Bertuglia and Vaio (2005) and Reggiani and Nijkamp (2009) relate the urban and regional science field to complexity science. Other recent books, typical of the genre, are those of Dorogovtsev and Mendes (2003), Nowak (2006), Desai and Kapral (2009), Phillipson and Schuster (2009), Barrat et al. (2008), Newman (2010) and Bianca and Bellomo (2011). These represent interesting presentations of the mathematics, but offer relatively few real system applications. There is an ongoing research task to monitor the relationships between the specific field of urban and regional science and the broader field of complexity science. The flows should be fruitful in both directions.

References

Barrat A, Berthélemy M, Vespigniani A (2008) Dynamical processes on complex networks. Cambridge University Press, Cambridge

Bertuglia CS, Vaio F (2005) Nonlinearity, chaos and complexity: the dynamics of natural and social systems. Oxford University Press, Oxford

Bianca C, Bellomo N (2011) Towards a mathematical theory of complex biological systems. World Scientific, Singapore

Casti JL (1995) Complexification. Abacus, London

Desai RC, Kapral R (2009) Dynamics of self-organised and self-assembled structures. Cambridge University Press, Cambridge

Dorogovtsev SN, Mendes JFF (2003) Evolution of networks. Oxford University Press, Oxford

Newman MEJ (2010) Networks. Oxford University Press, Oxford

Nowak AA (2006) Evolutionary dynamics. The Belknap Press of Harvard University Press, Cambridge

Phillipson PE, Schuster P (2009) Modelling by nonlinear differential equations. World Scientific, Singapore

Medda F, Nijkamp P, Reitveld P (2009) A morphogenetic perspective on spatial complexity: transport costs and urban shapes. In: Reggiani A, Nijkamp P (eds) Complexity and spatial networks. Springer, Berlin, pp 51–60

Further Reading

Holland JH (1995) Hidden order: how adaptation builds complexity. Addison-Wesley, Reading

Holland JH (1998) Emergence. Addison-Wesley, Reading

Wilson AG (2000) Complex spatial systems. Prentice Hall, New Jersey

Reggiani A, Nijkamp P (eds) (2009) Complexity and spatial networks. Springer, Berlin

Chapter 9
Research Challenges

9.1 Introduction

In the 'principles' presented in previous chapters, the argument has been under-
pinned by the STM approach: define the system of interest, think about the
underlying theory and then assemble appropriate methods for model building.
In this chapter, looking ahead to research challenges, and given that we have seen
many examples of systems, we change the order to TMS. A number of the
problems are rooted in past work in terms of opportunities missed or failures to
integrate different perspectives. Eight sections follow:

- expanding the conceptual tool kit
- expanding the methods tool kit
- information systems-1: representations and scales
- modelling challenges-1: understanding nonlinear systems
- modelling challenges-2: extending the range
- planning with models
- information systems-2: grand designs
- concluding comments

Interspersed with these are 13 examples which illustrate either current research
or future challenges; in each case, we focus on the main idea and provide refer-
ences to facilitate follow up and further work.

9.2 Expanding the Conceptual Tool Kit

The major research challenge is to move towards true interdisciplinarity, especially connecting economics to other modelling stream of work; and to an extent,
there is a similar argument for the rest of the social sciences. Why is it such a
problem? Most obviously, there are too many disciplinary silos. Economists in

A. Wilson, *The Science of Cities and Regions*, SpringerBriefs in Geography,
DOI: 10.1007/978-94-007-2266-8_9, © The Author(s) 2012

particular find it very difficult to connect to, even to recognise, other relevant approaches. Disciplines are powerful social coalitions, what Becher (1989) called 'academic tribes'. It is also a genuinely difficult task. A seemingly elementary starting point would be to take key concepts from economics and to bring them into, for example, BLV models. These might include demand, utility functions, production functions, profit, prices and rents. This has indeed been done in a rudimentary but illustrative way and an example will follow shortly. However, what has been done is partial. One difficulty is related to system representation and in particular, the use of discrete versus continuous space. We have argued that discrete representations facilitate the mathematics. Economists still mostly use continuous space. The example that follows shows what can be achieved: it is Herbert's and Stevens' (1960) reworking of Alonso (1960) in a discrete spatial representation. Alonso's work was in a more traditional continuous space mode and was in turn a reworking of von Thunen's thesis for a modern context. Herbert's and Stevens' model was then be reworked by Senior and Wilson (1974) by adding optimal dispersion via an entropy function. This demonstrates in a simple way how the different modelling styles can be reconciled. The research challenge is that this has not been followed through!

9.2.1 Example-1: Alonso–Herbert–Stevens–Senior–Wilson

Alonso's (1960, 1964) achievement was to show how von Thunen's (1826) bid rent concept could be applied in an urban setting. Herbert and Stevens (1960) shifted the model from continuous space to discrete space and set it up in mathematical programming form. They showed the duality between objective functions and constraints in a mathematical programming formulation. Senior and Wilson (1974) added an entropy function to represent an imperfect market. The research challenges, specifically, are: (1) how to bring about the entropy transformation for other economic models; (2) how to make maximum use of duality.

We now present the Herbert–Stevens model, modified slightly to facilitate the addition of an entropy function at a later stage. Define a set of spatial zones $\{i\}$, and population of w-income people in type k houses in zone i as T_i^{kw}. Let the land area be L_i, q^{kw}, the amount of land consumed by a type k house occupied by w-income people and p_i^{kw} the price of such a house in i. The b^{kw} is taken as the rent a w-income household would bid for a type k house. On the assumption that the land owner achieves a surplus of $b^{kw} - p_i^{kw}$, Herbert and Stevens argue that the optimum T_i^{kw} is derived from

$$\text{Max } B = \sum_{ikw} \left(b^{kw} - p_i^{kw} \right) T_i^{kw} \tag{9.1}$$

such that

$$\sum_{kw} T_i^{kw} q^{kw} \leq L_i \tag{9.2}$$

and

$$\sum_{ik} T_i^{kw} = P^w \tag{9.3}$$

The dual programme is

$$\text{Min } Z\{\alpha, v\} = \sum_i \alpha_i L_i - \Sigma_w v^w P^w \tag{9.4}$$

such that

$$\alpha_i q^{kw} - v^w > b^{kw} - p_i^{kw} \tag{9.5}$$

and

$$\alpha_i \geq 0 \tag{9.6}$$

As with all dual variables, they can be interpreted as shadow prices and α_i in this case functions as the rent per unit area commanded by the land owner in addition to the housing cost. Hence

$$r_i^k = q^{kw}\alpha_i + p_i^{kw} \tag{9.7}$$

is the rent actually paid.

This is a residential location model equivalent of the transportation problem of linear programming. In reality, this would produce too few non-zero T_i^{kw} and we can solve this problem by adding an entropy function. In illustrating this, we can also bring the model into line with the same aspect of the Lowry model (Chap. 1) by relating residential location to journey to work (via E_j^w, the number of type w jobs in j) and by introducing the housing stock explicitly as H_i^k. Then this slightly modified and disaggregated Herbert–Stevens model would be

$$\text{Max } B = \sum_{ijkw} T_{ij}^{kw}\left(b_{ij}^{kw} - c_{ij}^w\right) \tag{9.8}$$

such that

$$\sum_{jw} T_{ij}^{kw} = H_i^k \tag{9.9}$$

and

$$\sum_{ik} T_{ij}^{kw} = E_j^w \tag{9.10}$$

However, we can now optimally 'disperse' the system and maximising an entropy function and turning (9.8) into a constraint, the problem becomes

$$\text{Max } S = -\left[\sum_{ijkw} T_{ij}^{kw} \log T_{ij}^{kw}\right]/\mu \tag{9.11}$$

which is maximised subject to (9.9) and (9.10) as before and

$$\sum_{ijkw} T_{ij}^{kw}\left(b_{ij}^{kw} - c_{ij}^{w}\right) = B \tag{9.12}$$

(with B now taking a sub-optimal value). The result is

$$T_{ij}^{kw} = A_i^k B_j^w H_i^k E_j^w \exp\left[\mu\left(b_{ij}^{kw} - c_{ij}^{w}\right)\right] \tag{9.13}$$

with

$$A_i^k = 1\bigg/\sum_j^w B_j^w E_j^w \exp\left[\mu\left(b_{ij}^{kw} - c_{ij}^{w}\right)\right] \tag{9.14a}$$

and

$$B_j^w = 1\bigg/\sum_{ik} A_i^k H_i^k \exp\left[\mu\left(b_{ij}^{kw} - c_{ij}^{w}\right)\right] \tag{9.14b}$$

There is the advantage in this case that we can derive an analytical solution.

We can now proceed by making hypotheses about b_{ij}^{kw}. Recall that In the case of the retail model, for example, the equivalent of the b-term is $\log W_j$.

9.2.2 Example-2: Fujita–Krugman–Venables

For the detail of this, the reader is referred to Fujita et al. (1999). Their argument illustrates the restrictions imposed by conventional economics and the problems of modelling using continuous space. They also use the idea of 'variety of goods' as a key element of the utility function which, essentially, is a way of treating 'goods' as a continuum. The result is some difficult mathematics which generates a model which seems hugely implausible. The research challenge is to translate these ideas into discrete space, a more plausible definition of goods—improving the demand estimates by avoiding 'variety'—and add an entropy function. This would also facilitate moving beyond monocentricity of the workplace.

Fujita et al. also seem curiously disconnected from the other stream of urban economics: that which seeks to introduce dispersion not via entropy, but directly into utility functions—the random utility method which generates logit models, which then relate to entropy-based models—and this provides another route to integration.

In concluding this section, we should note that there are many other ways of expanding the conceptual tool kit. A continuing objective must be always to seek to apply the full tool kit to a particular problem and to integrate where possible. At the very least, an 'understanding' should be sought of what the different possible approaches can offer.

9.3   Expanding the Methods Tool Kit

We have seen that the initial research challenges on conceptual tools are through integration and this is also the case with methodological tools. The first such integration challenge is to fully relate mathematical and statistical approaches. At a philosophical level, these represent the hypothetico-deductive approach versus the inductive one: postulating a model and testing it against data against 'learning from the data' directly. Again, we find silos. Perhaps the sharpest distinction is revealed within economics between econometricians and mathematical economists. The approaches do ultimately come together, of course: mathematical models have to be calibrated and so 'maximum likelihood' statistical approaches kick in; and entropy-maximising models are, as we have seen, essentially Bayesian. Indeed, Berzins and Wilson (2003) showed that entropy models could be related directly to Fisher information. However, we should emphasise that purely statistical models are usually more restrictive both in format and in the assumptions that have to be satisfied before they are 'valid'. We might conjecture, for example, that mathematics handles nonlinearities better than statistics (There is a separate section on nonlinear systems below which indirectly, at least, offers evidence for this.).

9.3.1   Example-3: A Maths-Stats Challenge

There is an interesting specific research challenge that involves linking mathematics and statistics that connects to the current interest in scale-free phenomena. When the retail structural dynamics model is run, it generates equilibrium distributions of retail centre size, $\{W_j\}$. When this distribution is plotted, the graph is very much like a power law. The challenge is to demonstrate mathematically that this should be so, at least approximately. This is seemingly very difficult because, as we have seen, the equations for $\{W_j\}$ are not amenable to analytical treatment as shown in Chap. 7. The relevant equation is repeated here for convenience:

$$\sum_i \left\{ e_i P_i W_j^\alpha \exp\left(-\beta c_{ij}\right) \Big/ \sum_k W_k^\alpha \exp\left(-\beta c_{ik}\right) \right\} = k_j W_j \qquad (9.15)$$

The challenge is to find a way of extracting from these equations a probability distribution $P(W)$—the probability of a centre being of size W, given Eq. 9.15. Presumably, if this conjecture works at all, the result should be independent of the initial conditions. It may be worth exploring the $\alpha = 1$ case if this makes the problem simpler.

The next set of challenges involves relating models at different scales. The hypotheses (theories) represented in mathematical models depend very much on system representation and scale—to be discussed in detail in the next section. BLV models are meso-scale; ABMs and cellular automata models are more micro and rules-based. The question is: what rules in a micro-scale model will produce the same results in a meso-scale model? This has been explored in Dearden and Wilson (2011), as discussed in Chap. 7, but has not been demonstrated mathematically.

There are many other mathematical challenges to be met. For example, the need to understand fully, and to take advantage of, the duality between objective functions and constraints; to formally understand how equilibria in different types of model are all rooted in Brouwer's (1910) fixed point theorem (and cf. Scarf 1973a, b); and to develop further the equivalences between different approaches. An under-exploited area for example is game theory and Nash equilibria. There are challenges of incorporating 'learning' and innovation in models—cf. Brian Arthur (2009) on 'technologies'—and this links, probably at best indirectly, to neural network models in which the models 'learn' from the data. Yee Leung (1997) produced a neural network model that was equivalent to a spatial interaction model.

The research challenge is to find more equivalences and thereby to expand the tool kit; ideas will come from trying to integrate.

9.4 Information Systems-1: Representations and Scales

There are research challenges in assembling information systems to underpin our research—partly because we operate on scales that vary from the micro (individuals, households, firms) to the macro (urban, regional, national, global) and also, a related issue, because we have so many potential categories to handle for each element of our systems of interest. We need to start with a top down view and define information systems that will embrace our research questions: system of interest, functioning and evolution; and chains of causality. This drives us in the direction of having as comprehensive a 'picture' as possible even if our research questions are more specific—because of the 'chains'. We then need to relate this to a bottom-up perspective: what data is available? Can missing 'data' be estimated in some way? We are (typically) interested in (starting at the finest scale so we can aggregate as appropriate):

- individuals, households and their characteristics
- firms and service organisations
- 'sectors' (of an economy e.g.)
- spatial units

This 'picture' would be repeated for different points in time, and for time intervals. 'Space' requires a zoning system (unless it is treated as continuous); ideally, we need borderless systems; otherwise, we have to introduce external zones to close the system. A problem will be that the data will have 'categories' that are not typically the ones we want and different sources will not be consistent and so a further research challenge is to use something like the entropy-maximising framework to force a best estimate of an appropriate and consistent data set.

9.4.1 Example-4: The 10^{13} Issue

Consider the following characterisations of variables (cf. Wilson 2007) for a system description with number of categories of each index in brackets:

population: age (7), gender (2), family structure (4), education level (4), occupation (7), income (6), car availability (2), residential location (100): $7 \times 2 \times 4 \times 4 \times 7 \times 6 \times 2 \times 100 = 2$ M (approx)
similarly, services, public services, manufacturing, housing, agriculture and land generate something of the order of 1 M each

The interaction arrays are then enormous: 5 at 2×1 M, so something like 10^{13} variables are needed for a complete system description. A 1000-zone system would imply 10^{15} and hence the need for aggregation and/or microsimulation.

The research challenge is to use this kind of algebraic system description to structure an underlying information system that can function at various levels of aggregation.

9.4.2 Example-5: Using Geodemographics in a BLV System

Given the 10^{13} challenge, it is worthwhile exploring other avenues of aggregation. A potentially effective way of doing this is through geodemographics; then a string of population characteristics, $\{n_1, n_2, n_3, n_4, ...\}$ can be collapsed into one, n. This approach has been explored for the higher education system by Singleton et al. (2011). This is an interesting illustration of what could be the beginnings of a new research area. It also illustrates a different kind of spatial system: 150 English local authorities as residential origins and 88 universities as destinations. The families of students, and hence it is assumed the students themselves, are classified into seven OAC supergroups. The model used illustrates another modelling task referred to briefly in Chap. 6: models that are partially constrained at the destination end, in this case, because some universities are 'selecting' and assumed to be constrained, while others are 'recruiting', and less so. The model used is:

$$S_{ij}^{fg} = A_i^{fg} O_i^{fg} (R_j)^{\alpha} K_j^{g} \exp\left(-\beta_{1i}^{f}\beta_{2ij}c_{ij}\right) \sum_i S_{ij} \leq K_j^{g} \qquad (9.16)$$

$$S_{ij}^{fg} = A_i^{fg} B_j^{g} O_i^{fg} K_j^{g} \exp\left(-\beta_{1i}^{f}\beta_{2ij}c_{ij}\right) \sum_i S_{ij} > K_j^{g} \qquad (9.17)$$

$$A_i^{fg} = \sum_j (R_j)^{\alpha} K_j^{g} \exp\left(-\beta_{1i}^{f}\beta_{2ij}c_{ij}\right) \sum_i S_{ij} \leq K_j^{g} \qquad (9.18)$$

$$A_i^{fg} = \sum_j B_j^{g} O_i^{fg} K_j^{g} \exp\left(-\beta_{1i}^{f}\beta_{2ij}c_{ij}\right) \sum_i S_{ij} > K_j^{g} \qquad (9.19)$$

$$B_j^{g} = \sum_i A_i^{fg} O_i^{fg} \exp\left(-\beta_{1i}^{f}\beta_{2ij}c_{ij}\right) \sum_i S_{ij} > K_j^{g} \qquad (9.20)$$

where
S_{ij}^{fg} the number of students from local authority i in OAC group f with attainment level g going to university j
O_i^{fg} the number of students from local authority i in OAC group f with attainment level g going to university
R_j the attractiveness of university j
K_j^{g} the capacity of university K for admitting students of attainment level g
β_{1i}^{fg} an element of the distance parameter relative to distance decay
β_{2ij} an element of the distance parameter—a device for relative attainment to enhance requirement
c_{ij} a measure of the 'distance' between i and j as a surrogate for impedance

In the singly-constrained model, there is no B_j coefficient. In the doubly-constrained model, both A_i and B_j coefficients are present and depend upon each other, so the equations are solved iteratively. Each university also has a composite attractiveness factor $R_j^{\alpha}K_j$ which is, of course, only used in the singly-constrained part of the model. The student capacity is used in the doubly-constrained model.

A research challenge, therefore, is to use geodemographics in this way with other systems, or more realistically with this one.

9.4.3 Example-6: Representing Multi-Level Systems

Consider the following problem, which arises in the study of migration but which contains a more general idea: how do we develop first an effective notation, and then an appropriate model, when we have two levels of spatial aggregation and constraints at each level—but we only have partial information at the finer-scale. In the migration case that illustrates this idea, we have a system of countries, labelled with capital letters such as I and J; and within each country, a system of smaller zones labelled, say, i and j. Then let T^{IJ} be the number of migrants from country I to country J in some time period, say t to $t + 1$ (but we will leave that implicit for ease of notation). Then we can denote by T_{ij}^{IJ} the number of migrants from region i in I to region j in J. For convenience of notation, we denote all the migration flows by T but the different numbers of subscripts and superscripts indicate the different levels. This notation implies that we number the finer-scale zones from $1,......n_I$ for country I rather than numbering them consecutively for the whole system.

At the fine-scale, data is available for the $\{T_{ij}^{II}\}$ array—that is the diagonal elements of the $\{T_{IJ}\}$ matrix—and, we assume, with whole country numbers for the rest of the matrix.

The row and column total are known for the $\{T_{ij}^{II}\}$ elements and also for the T^{IJ} inter-country levels. Let these be $\{M_i^I\}$ and $\{N_j^I\}$ and $\{O^I\}$ and $\{D^J\}$ respectively so that

$$\sum_j T_{ij}^{II} = M_i^I \qquad (9.21)$$

$$\sum_i T_{ij}^{II} = N_j^I \qquad (9.22)$$

$$\sum_J T^{IJ} = O^I \qquad (9.23)$$

$$\sum_I T^{IJ} = D^J \qquad (9.24)$$

A decision would have to be made at this point as to whether O^I and D^J include the $\{T_{ij}^{II}\}$ terms—that is the intra-country flows. The formulation of the above sums imply that the intra-zonal totals are included and that in (9.23) and (9.24).

$$T^{II} = \sum_{ij} T_{ij}^{II} \qquad (9.25)$$

We should also assume that the data is cleaned up to an extent that

$$\sum_i M_i^I = \sum_j N_j^I = T^{II} = \sum_{ij} T_i^{II} \qquad (9.26)$$

The formulation thus far implies that we are not seeking to model flows at the fine-scale level from each country, I, to and from other countries, J, $J \neq I$. So we have to introduce another element of notation: T_{iJ}^I and T_{Ji}^I are, respectively, the out-migration flow from zone i in country I to country J ($\neq I$) and the in-migration flow from country J ($\neq I$) to zone i in country I. T^{IJ} in (3) and (4), for $I \neq J$, would then be given by

$$T^{IJ} = \sum_{i \varepsilon I} T^I_{ij} = \sum_{i \varepsilon I} T^I_{Ji} \tag{9.27}$$

Once these notational issues are settled—in effect ensuring a good system description—then we can consider the variety of models that could be built. The equations such as (9.21)–(9.24) also provide some of the core constraint equations for entropy-maximising models. The possibilities are:

(1) Model the fine-scale flows within each country separately—that is model $\{T^{II}_{ij}\}$ (in which case I simply functions as a label for each country model). Equations 9.21 and 9.22 would be the accounting/constraint equations.
(2) Model the inter-country flows $\{T^{IJ}\}$, separately. Equations 9.23 and 9.24 would be the accounting equations.
(3) Model fine-scale flows within each I, $\{T^{II}_{ij}\}$; zonal flows in and out of each I to each J ($\neq I$), $\{T^I_{ij}\}$ and $\{T^I_{Ji}\}$; and inter-country flows $\{T^{IJ}\}$, $I \neq J$. In this case the full set of Eqs. 9.21–9.24 would hold as accounting/constraint equations and these would have to be written explicitly in terms of the T^I_{ij} and T^I_{Ji} variables.
(4) Model the full array of fine-scale zones, $\{T^{II}_{ij}\}$. The same considerations on accounting equations and variables hold as for model (3).

If all the accounting Eqs. 9.21–9.24 are deployed, this leads to the construction of doubly-constrained models for which the main task would be to identify impedance functions, associated generalised costs, c_{ij}, and the model parameter values through model calibration. Model (1) is the most straightforward and would produce something like:

$$T_{ij} = A_i B_j M_i N_j \exp\left(-\beta_i c_{ij}\right) \tag{9.28}$$

The II country labels have been dropped for convenience. This model would be run separately for each country.

The inter-country model (2) would be

$$T^{IJ} = A^I B^J O^I D^J \exp(-\mu_I c_{IJ}) \tag{9.29}$$

With this kind of formulation, the best way to introduce, for example, push and pull factors to enrich the models, would be to introduce submodels for M^I_i, N^I_j, O^I and D^J. Formally

$$M^I_i = M^I_i\left(X^{1I}_i, X^{2I}_i, \ldots\ldots\right) \tag{9.30}$$

$$N^I_j = N^I_j\left(Z^{1I}_j, Z^{2I}_j, \ldots\ldots\right) \tag{9.31}$$

$$O^I = O^I(X^{1I}, X^{2I}, \ldots\ldots\ldots) \tag{9.32}$$

$$D^J = D^J(Z^{1J}, Z^{2J}, \ldots\ldots\ldots) \tag{9.33}$$

Model (3) is potentially much more complicated; and model (4) is simpler in principle, but probably unrealistic at the present time. Model (3) to look something like the following.

$$T_{ij}^{II} = A_i^I B_j^I M_i^I N_j^I \exp\left(-\beta_i^I c_{ij}^I\right) \tag{9.34}$$

$$T_{iJ}^I = A_i^I B_j^I M_i^I D^J \exp\left(-\beta_i^I c_{ij}^I\right), \quad I \neq J, \tag{9.35}$$

$$T_{Jj}^I = A^I B_j^I O^I N_j^I \exp\left(-\beta_i^I c_{ij}^I\right), \quad I \neq J \tag{9.36}$$

$$T^{IJ} = A^I B^J O^I D^J \exp(-\mu_I c_{IJ}), \quad I \neq J \tag{9.37}$$

There could be submodels for the M, N, O, D variables as in Eqs. 9.30–9.33.

9.5 Modelling Challenges-1: Understanding Nonlinear Systems

We should recall the main features of most of our systems of interest: we are modelling in high-dimensional spaces with nonlinear mechanisms such as positive returns to scale. As we have seen these features generate multiple equilibria, phase changes and path dependence. The first research challenge is to identify explicitly more of these features in a variety of systems of interest.

9.5.1 Example-7: Unpicking the Mechanisms of Phase Change

In modelling high-dimensional systems with high levels of complexity through interdependence, it is very difficult to get an analytical grip on the process of evolution. The starting point was offered by Harris and Wilson (1978) and further developed in Dearden and Wilson (2011).

The original idea is shown in the graphs in Fig. 9.1—originally introduced and discussed in Chap. 7 but repeated here for convenience. This shows a revenue plot for a retail centre and two versions of the cost line—one intersecting the curve, one not. In the case of the intersecting line, the two end points are stable—thus illustrating multiple equilibria because both are possible—and the central inter-section is unstable. What is demonstrated in Dearden and Wilson is that a study of

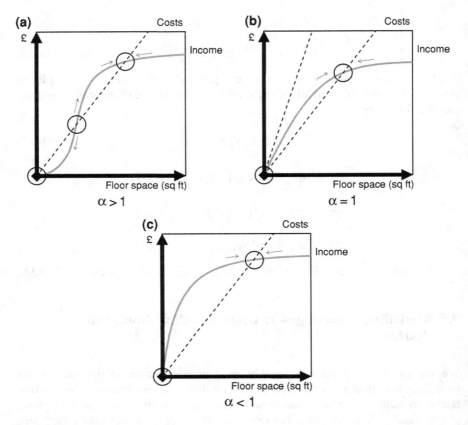

Fig. 9.1 Revenue-costs zone graph

initial values for each zone will reveal new insights about the development path. For example, for a particular zone, if the initial value is below the unstable point, then development in that zone will be impossible. The research challenge is to carry out this kind of analysis for a greater variety of systems that are generating structures—e.g. reaction–diffusion systems—see below.

Even in the retail case, there is scope for doing this with improved models of the elements: consumer behaviour and production functions for example. The current models are very crude in this respect—operating on a 'proof of concept' basis. Recall

$$\Delta W_j = \varepsilon [D_j - KW_j] W_j \tag{9.38}$$

It would be interesting to focus on the KW_j term for example—introducing a better production function. Note than then the straight line representing cost would become a curve in Fig. 9.1 though the essence of the argument about intersection and stable and unstable points would be the same.

9.5.2 Example-8: An Extended Dynamic Model

In the dynamical models presented so far, and particularly in the retail demonstrator—we have focused on the retail centre sizes. There are possibilities of a new way of integrating with economic modelling by explicitly introducing the prices of goods and a geographically varying rent—as in the Herbert and Stevens in Example 1 above. Let p_j be an average price of goods sold in j and let r_j be the unit rent in j. Then p_i can be built into the e_i to improve the demand function, and as $p_j^{-\gamma}$ say, as a factor in the attractiveness function and K becomes $(K + r_j)$. We can then have three dynamics equations:

$$\Delta W_j = \varepsilon\left(D_j - KW_j\right)W_j \tag{9.39}$$

$$\Delta p_j = \mu\left(D_j - KW_j\right)p_j \tag{9.40}$$

$$\Delta r_j = v\left(D_j - KW_j\right)r_j \tag{9.41}$$

and the sizes of ε, μ and v will determine the relative importance of the three different drivers. This, of course, would be the basis of new empirical research.

9.5.3 Example-9: To Explore and Integrate the Range of Models that Generate Structure

In Chap. 3, we showed how the models of Richardson, Lotka and Volterra, Harris and Wilson and the Turing/Medda reaction diffusion model argument could all generate emergent structure. There must be other set of simultaneous differential or difference equations that could do this. One specific question is: are these all topologically related at some deeper level? But the research challenge is to locate these 'other' systems. This is to some extent putting the cart before the horse—looking at a 'method' before we have the system or the theory, but it may be a way of stimulating new thinking.

9.5.4 Example-10: 'DNA' and Typologies of Cities

The evolution of a nonlinear dynamical system is very dependent, as we have seen, on the initial conditions and path dependence can be thought of as a sequence of such initial conditions. Since these are connected to 'slow' dynamics, these initial conditions can be thought of as the 'DNA' of the system. Are there then groups of characteristic 'initial conditions' that can be thought of as 'genes'? If so, this opens up the possibility of characterising typologies of cities in relation to 'gene' structure. There is then a possible link to remote sensing which could be thought of as the X-ray crystallography of urban analysis. The first research challenge is then to explore the potential range of technologies and a second is to find real examples, possibly historical—cf. Wilson (2010a, b).

9.6 Modelling Challenges-2: Extending the Range

Another set of research challenges come from extending the modelling range. There are three main directions of travel here: engaging with new disciplines; building new models; and responding to planning or business questions. We discuss each in turn.

The possible new disciplines, some already explored in a preliminary way, include history (and archaeology) for example. This is not a new idea.

"This work is composed of Geographie (which is a description of the known earth and the parts thereof) and Historie, which is the eye of the world. These two goe inseparably together, as it were hand in hand..... and are like two Sisters entirely loving each other, and cannot without pitie be divided."

Preface to the first English Edition of Mercator's *Atlas* (1636) by Henry Hexham, translator; quoted by Maurice Beresford (1957) *History on the ground*, Lutterworth Press, London

Why did it take so long?!

We can now add political science via the already cited Richardson and Epstein; ecology, where we have shown how to add 'space'—so the modelling of ant colonies and the like could prove to be a rich field; epidemiology, again, as we have seen, adding space; physics, making more use of the formalism—the thermodynamics of the city etc. (Wilson 2009).

The next step is to explore new models. There is still plenty of scope for doing this—many opportunities arising from the argument so far. But let us consider briefly three further examples. We could change scale and model pedestrians in shopping centres; this would demand a time-sequenced spatial interaction model. Or we pursue two new approaches to the emergence of hierarchies via the next two examples. If these avenues of research prove fruitful, they could then be extended into other areas such as Lowry-type general models. A third example is then offered showing how different elements of the modellers' tool kit can be used to assemble a prototype global dynamics model.

9.6.1 Example-11: Hierarchical Retail Models

In the Harris–Wilson model, the 'hierarchy' is represented by a continuous spectrum within the $\{W_j\}$ vector. A research challenge is to make the hierarchy more explicit through representing explicitly the production functions at different levels of a hierarchy?

Suppose we have two hierarchical levels, labelled by $h = 1, 2$, the second being the 'supermarket'. Let W_j^h be the size of h in zone j. Let K^h be the cost per unit, say annually. Let e_i be the unit expenditure by the population of i and P_i the population. Let S_{ij}^h be the set of flows, measured in money units. Then a suitable model could be

$$S_{ij}^h = A_i e_i P_i \left(W_j^h\right)^{\alpha h} \exp\left(-\beta^h c_{ij}\right) \qquad (9.42)$$

where

$$A_i = 1 / \sum_k^h \left(W_k^h\right)^{\alpha h} \exp\left(-\beta^h c_{ij}\right) \qquad (9.43)$$

to ensure that

$$\sum_{jh} S_{ij}^h = e_i P_i. \qquad (9.44)$$

The retail flows from zone i to zone j will be split between $h = 1$ and $h = 2$ according to the relative sizes of the Ws and the parameters α^h and β^h. The dynamics, in difference equation form, can be taken as

$$\Delta W_j^h = \varepsilon \left[D_j^h - K^h W_j^h\right] W_j^h \qquad (9.45)$$

where

$$D_j^h = \sum_i S_{ij}^h \qquad (9.46)$$

We will assume that $K^{(2)}$ is greater than $K^{(1)}$ because of scale economies. However, these economies will only be achieved if W_j^h is greater than some minimum size. Say

$$W_j^{(2)} \geq W_j^{(2)\min} \qquad (9.47)$$

and we expect that $W_j^{(1)}$ is very much smaller than $W_j^{(2)\min}$.

The model therefore consists of Eqs. 9.42–9.47 run in sequence. In the first, the equilibrium solutions to

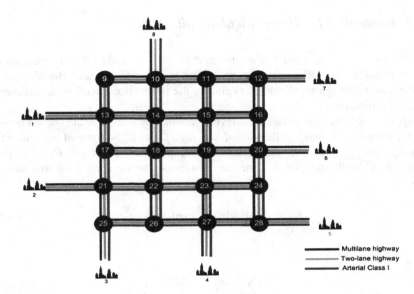

Fig. 9.2 The elements of an hierarchical network

$$D_j^h = K^h W_j^h \tag{9.48}$$

should be explored. And then assumptions could be made about the trajectories of exogenous variables so that the dynamics could also be explored. It would be quite likely that for certain sets of values of exogenous variables, there would be no $W_j^{(2)} > 0$ and as, say, income increased, we could model the emergence of supermarkets. This may be interesting in its own right, but as noted at the outset, the current motivation is to point the way to modelling the evolution of transport networks.

9.6.2 Example-12: The Evolution of Networks

It is very much harder to model network evolution than, say, that of retail structure. A way is suggested in the reference above (with Example 10), essentially based on the Fig. 9.2.

This argument follows that in Wilson (1983) but aims first to articulate the simplest transport model that would demonstrate the desired properties and into which we can insert explicit hierarchies. The 1983 paper contains a model that is rich is detail, but also ends with a formulation of a simpler model, but one which does not have explicit hierarchies. In the simpler model of that paper, spider networks are deployed to seek to avoid the complexities of detailed real networks (which had been spelled out earlier in that paper). We use the same method here: spider networks are built by connecting nearby zone centroids with notional links.

Let $\{i\}$ be a set of zones and these elements can also stand as names for zone centroids. We then elaborate the Chap. 6 account of assignment by introducing explicit hierarchical levels, letting $h = 1, 2, 3$ label the three levels of hierarchy mentioned in the introduction—minor, arterial, motorway. We will also use j, u and v as centroid labels. For example (u, v, h), may be a link on a route from i to j on the spider network at level h. $(u, v)\ \varepsilon\ R_{ij}^{\min}$ is the set of links that make up the best route from i to j. This may involve a mix of links of different levels. Let Γ^h be a measure of the annual cost per unit of a link of level h—including an annualised capital cost—and let ρ_{uv} be the 'length' of link (u, v). Let c_{ij} be the generalised cost of travel from i to j as perceived by consumers, and let $\{\gamma_{uv}^h\}$ be the set of link costs on the level h link. So

$$c_{ij} = \sum_{(u, v, h)\ \varepsilon\ Rij\min} \gamma_{uv}^h \tag{9.49}$$

If O_i and D_j are the total number of origins in zone i and destinations in zone j, and T_{ij} is the flow between i and j, then the standard doubly-constrained spatial interaction model in the usual way would be

$$T_{ij} = A_i B_j O_i D_j \exp(-\beta c_{ij}) \tag{9.50}$$

with

$$A_i = 1/\sum_k B_k D_k \exp(-\beta c_{ik}) \tag{9.51}$$

and

$$B_j = 1/\sum_k A_k O_k \exp(-\beta c_{ki}) \tag{9.52}$$

To model congestion, we need to know the flows on each link. Let q_{uv}^h be the flow on link (u, v, h) and let Q_{uv}^h be the set of origin–destination pairs at level h that use the (u, v, h) link. Then

$$q_{uv}^h = \sum_{ij\varepsilon Quvh} T_{ij} \tag{9.53}$$

We then want γ_{uv}^h to be a function of the flow:

$$\gamma_{uv}^h = \gamma_{uv}^h\left(q_{uv}^h\right) \tag{9.54}$$

For this simple demonstration model, we first assume that this function is the same for each link in a particular level of the hierarchy. This can be derived from standard speed-flow relationships. Then the equilibrium position of the network can be obtained by following the sequence of Eqs. 9.49–9.54 and iterating. This will depend very much on the initial conditions and in particular, on the links that are in place at higher levels in the hierarchy. These initial conditions represent the 'DNA' of the system at this time—cf. Wilson (2010). The dynamics, in this case can be thought of as the addition of higher level links into the network, within a budget, to minimise an objective function—say, consumers' surplus, or for the sake of simplicity, total travel (generalised) costs measured in money units. This would be a mathematical programme of the form

$$\text{Min } C = \sum_{ij} T_{ij} c_{ij} \tag{9.55}$$

subject to (9.49)–(9.54) and

$$\sum_{uv \, s.t. \, \gamma uvh(t)=\infty \Gamma^h} \rho_{uv} = \Gamma(t+1) \tag{9.56}$$

The summation in (9.56) is over possible links that do not exist at time t, so that new links can be added up to an incremental budget spend of $\Gamma(t+1)$.

The dynamic model would then take the form: run (9.55)–(9.56), insert the new γ_{uv}^h into (9.49)–(9.54) and iterate as usual (as an inner iteration) and then begin the outer iteration again with a repeat of (9.55)–(9.56). This is easy to formulate conceptually. The mathematical programme (9.55)–(9.56) may be difficult to handle in practice.

This leads to the possibility of a second dynamic model in which each link at each level of the hierarchy is given a capacity, x_{uv}^h and the link costs then become functions of this capacity as well as the flows:

$$\gamma_{uv}^h = \gamma_{uv}^h \left(q_{uv}^h, x_{uv}^h \right) \tag{9.57}$$

We could then think of a difference equation in Δx_{uv}^h. Suppose we take γ_{uv}^h as a measure of congestion—that is if the unit generalised cost on a link is high, we attribute this to congestion (There will still be the problem of dealing with non-existent links where this is formally infinite.). We could control the costs in this case by increasing Γ^h in an iterative cycle to ensure that a budget constraint is met. This suggests

$$\Delta x_{uv}^h = \varepsilon^h \left[\gamma_{uv}^h - \gamma \right], \quad \left[\gamma_{uv}^h - \gamma \right] > 0, \quad \Delta x_{uv}^h = 0 \text{ otherwise} \tag{9.58}$$

where γ is taken as a threshold and ε is calculated to ensure that the budget constraint holds for this time period:

$$\sum_{uv \, s.t. \, \gamma uv(t)=\Gamma^h} \rho_{uv} = \Gamma(t+1) \tag{9.59}$$

$x_{uv}^h(t+1)$ would then be fed back into (9.49)–(9.55)—with (9.58) replacing (9.55).

The challenge is to build this kind of model and explore the dynamics! A demonstration model is offered in Pagliara et al. (2011).

9.6.3 Example-13: The Fry–Wilson 'Global Dynamics' Demonstration Model

In the volatile world in which we live, any model-based understanding of 'global dynamics' will be an asset. This example is taken from a new project which aims to model trade and migration in relation to development aid and security. There would obviously be a role for inter-regional input–output modelling—or, in terms of this example, a search for an approximation because at present it is technically impossible, if only because of data shortages, to have a full model for a 220-country system. What would the 'approximation' look like? It has to be sufficiently realistic to represent the most important issues and to be able to differentiate types of country, from the very poor to the very rich. Fry and Wilson (2011), building on Wilson (2010), have sought to break into this agenda with a three sector international trade model at the heart of which is a trade flow spatial interaction model of the form:

$$Y_{ij} = Z_j q_i \exp\left[-\beta\left(\vartheta d_{ij} + \varphi_i\right)\right] / \sum_k q_k \exp\left[-\beta\left(\vartheta d_{kj} + \varphi_k\right)\right] \qquad (9.60)$$

The Y_{ij} are the flows, Z_i, total production, q_j demand and d_{ij} the distance between I and j. θ is a unit cost of transport and φ_k the price at k.

9.7 Planning with Models

Ashby's 'law of requisite variety' makes an interesting starting point. It argues that for a system that is to be 'controlled' the control system must have at least as much 'variety'—i.e. complexity—as the system of interest. This offers huge insight. What models offer the planner is some of that variety, especially if they can be used in 'What if?' forecasting or plan/scenanrio testing modes [This links to Friston's (2003) ideas of 'representing the environment in the brain—using the notion of 'free energy' which resonates with the ideas presented here.]. We have used the retail model as a key example and noted that this is extensively used by retailers—'retailers' in the broadest sense—banks, car manufacturers in relation to their dealerships, utilities' companies and oil companies in relation to garage outlets. The utility that can be gained from this kind of intelligence has not, in general, been similarly realised in relation to public services, with the exception of the transport sectors—and probably defence, though most of that work is secret. This means that there is an enormous range of research opportunities to demonstrate what could be gained. Indeed, it can be argued that new insights can be achieved in relation to the most difficult problems facing governments—national and local. We consider a number of examples in turn to show what is in principle possible.

We have seen that the kinds of models described in this book have been widely applied in the transport and retail sectors and that comprehensive models have been developed which can be applied in planning—though this has not been widespread. There are a number of other sectors in which these kinds of models could be systematically applied. The development of these would make good research projects! Examples follow.

(1) In health planning, it is possible to use spatial interaction models to predict patient flows to different kinds of facilities. These could then be used to underpin the planning of the size and location of facilities at primary, secondary and tertiary levels. There are subsidiary research questions on the effectiveness of care by facility size.

(2) A similar argument applies in education whether for primary, secondary, further or higher. It is possible to explore the basis of widening access policies for example.

(3) Different kinds of models could be built to address issues relating to policing and the criminal justice system (CJS). Crime rates in small areas could be linked, for example to the location of police resources—whether 'stations' or mobile. In the case of the CJS, an interesting model-based analysis could be carried out in relation to prisons. The locational challenge is to locate prisoners, in many cases, as near as possible to their homes but subject to constraints that the prisons they are in have the courses available to facilitate their rehabilitation.

(4) We have already indicated in Chap. 2 on trade, in Chap. 3 on wars and security and in Example 13 above, that models can be deployed on the global scale and this is relevant for a variety of government departments: foreign, defence, trade and development aid for example.

(5) Within a comprehensive model that identified the importance of income from work or other sources such as pensions, it would be possible to models the impacts of different employment, benefits and pensions policies.

(6) It would be possible to build spatial interaction models relating usage of cultural facilities and then to use related indicators, such as accessibilities and effective delivery, to analyse support policies.

See Wilson (2008b) for a more detailed analysis of possibilities.

9.8 Information Systems-2: Grand Designs

We now take the argument about information systems a stage further. Assume we can build an information system along the lines of Sect. 9.3, taking into account the difficulties raised in Example 4. If such a system can be aggregated in a variety of ways, and if it can be intelligently 'searched', then, among other things, we should have created an atlas for the e-age with access to many millions of maps. Such an information system would contain data, model outputs, plans, issues, foresight; and through model runs would be able to generate a variety of indicators. In effect, we would have a model-based intelligent GIS.

The first associated research challenge is to design and build the architecture of such a system. This is non-trivial and would be the basis for best practice in model-based analysis and planning. The second challenge is to make more effective use of indicators to represent the state of the system of interest—building on, for example, Williams (1977) and Clarke and Wilson (1987a, b).

9.9 Concluding Comments

Many of the tools outlined here have been available for a long time, but they are not usually seen as a comprehensive package and there is a danger that there is much re-inventing of the wheel. There is a great variety of past experience to build on. How do we achieve best practice across so many potential fields? How do we emulate the transport modellers? The 'general urban' modellers and the input–output modellers come close. Are we about to re-do the GUM with the tools of nonlinear systems analysis? Universities can be the research front-line in these endeavours–on a 'proof of concept' basis—and then Government agencies with serious funding available should follow!

References

Alonso W (1960) A theory of the urban land market. Pap Reg Sci Assoc 6:149–157
Alonso W (1964) Location and land use. Harvard University Press, Cambridge
Arthur WB (2009) The nature of technology. Allen Lane, London
Becher T (1989) Academic tribes and territories: intellectual inquiry and the culture of disciplines. Open University Press, Milton Keynes
Berzins M, Wilson AG (2003) Spatial interaction models and fisher information: a new calibration algorithm. Environ Plan A 35:2161–2176
Brouwer LEJ (1910) Uber eineindeutige stige transformationen von flachen in sich. Math Ann 67:176–180
Clarke GP, Wilson AG (1987a) Performance indicators and model-based planning I: the indicator movement and the possibilities for urban planning. Sistemi Urbani 2:79–123
Clarke GP, Wilson AG (1987b) Performance indicators and model-based planning II: model-based approaches. Sistemi Urbani 9:138–165
Friston K (2003) Learning and inference in the brain. Neur Netw 16:1325–1352
Fujita M, Krugman P, Venables AJ (1999) The spatial economy: cities, regions and international trade. MIT Press, Cambridge
Harris B, Wilson AG (1978) Equilibrium values and dynamics of attractiveness terms in production-constrained spatial-interaction models. Environ Plan A 10:371–388
Herbert DJ, Stevens BH (1960) A model for the distribution of residential activity in an urban area. J Reg Sci 2:21–36

Leung Y (1997) Intelligent spatial decision support systems. Springer, Berlin

Pagliara F, Wilson A, de Martinis V (2011) The evolution of hierarchical transport networks: a demonstration model. Working Paper 169, Centre for advanced spatial analysis, University College London

Scarf H (1973a) The computation of economic equilibria. Yale University Press, New Haven

Scarf H (1973b) Fixed-point theorems and economic analysis. Am Scientist 71:289–296

Senior ML, Wilson AG (1974) Explorations and syntheses of linear programming and spatial interaction models of residential location. Geog Anal 6:209–238

Singleton AD, Wilson AG, O'Brien O (2011) Geodemographics and spatial interaction: an integrated model for higher education. J Geog Syst

von Thunen JH (1826) Der isolierte staat in beziehung auf landwirtschaft und nationalokonomie. Gustav Fisher, Stuttgart. English edition: von Thunen JH (1966) The isolated state (trans: Wartenburg CM). Oxford University Press, Oxford

Williams HCWL (1977) On the formation of travel demand models and economic evaluation measures of user benefit. Environ Plan A 9:285–344

Wilson AG (1983) Transport and the evolution of urban spatial structure. In Atti delle Giornate di Lavoro 1983 Guida Editori, Naples, pp 17–27

Wilson AG (2007) A general representation for urban and regional models. Comput Environ Urban Syst 31:148–161

Wilson AG (2008a) Boltzmann, Lotka and Volterra and spatial structural evolution: an integrated methodology for some dynamical systems. J R Soc Interface 5:865–871. doi:10.1098/rsif.2007.1288

Wilson AG (2009) The 'thermodynamics' of the city. In: Reggiani A, Nijkamp P (eds) Complexity and spatial networks. Springer, Berlin, pp 11–31

Wilson AG (2010a) Knowledge power. Routledge, London and New York

Wilson AG (2010b) The general urban model: retrospect and prospect. Pap Reg Sci Assoc 89:27–42

Further Reading

Birkin M, Clarke GP, Clarke M, Wilson AG (1996) Intelligent GIS: location decisions and strategic planning. Geoinformation International, Cambridge

Birkin M, Clarke GP, Clarke M (2002) Retail geography and intelligent network planning. Wiley, Chichester

Dearden J, Wilson AG (2011) A framework for exploring urban retail discontinuities. Geog Anal

Wilson AG (2008b) Science and the city. Environ Plan A 40:2800–2808